The
Pattern and Function
Connection

Brad S. Fulton
Bill Lombard

Key Curriculum Press
Innovators in Mathematics Education

Project Editors: Christian Aviles-Scott, Mary Jo Cittadino

Editorial Assistants: Heather Dever, Jason Taylor

Teacher Reviewers:
Ed Bartel, David A. Brown Middle School, Wildomar, California
Claudia Burris, La Mesa–Spring Valley School District, La Mesa, California
Judy Ferrari, Pleasant Hill Junior High School, Pleasant Hill, Oregon
Julie Moshier, Tulare Western High School, Tulare, California
Magda Noffal-Perkins, Matilija Junior High School, Ojai, California
Vicki A. Vierra, E. O. Green Junior High School, Oxnard, California

Mathematics Reviewer: Larry Copes

Production Editor: Kristin Ferraioli

Copy Editor: Margaret Moore

Production and Manufacturing Manager: Diana Jean Parks

Production Coordinator: Laurel Roth Patton

Text Designer and Compositor: Andrea Reider

Art Editor: Jason Luz

Technical Artist: Derrick Hardy

Art and Design Coordinator: Caroline Ayres

Cover Designer: Burl Sloan

Prepress and Printing: Versa Press, Inc.

Executive Editor: Casey FitzSimons

Publisher: Steven Rasmussen

Key Curriculum Press
1150 65th Street
Emeryville, CA 94608
510-595-7000
editorial@keypress.com
http://www.keypress.com

Printed in the United States of America

10 9 8 7 6 5

ISBN 978-1-55953-395-9

Contents

Introduction

The Pattern and Function Connection is a book written by teachers, to teachers, for students. We know how busy you are. In addition to the literally hundreds of duties and responsibilities you already have, you may also be teaching a class without a suitable textbook or other source of curriculum. You simply don't have the time to find all the material you need. That's where we hope to help.

With *The Pattern and Function Connection* you and your students are about to discover a progressively paced set of activities that can be used either individually or as an excellent introductory unit. Together as a class you will learn about linear algebraic functions. Your students will discover the interconnectedness between a visual pattern, a T-table, a graph, and a formula and they will become proficient at using both T-tables and graphs for problem solving. Best of all, your students will develop an awareness of the symbolic world of algebra while becoming adept at working with it.

This book is based on specific and important goals that we think you'll appreciate. First of all, as teachers ourselves, we know first hand that many texts and units are either too symbolic or solely concrete and manipulative based. Pre-algebra and algebra students are students in transition. They learn from the concrete, but they are capable of moving into the abstract. *The Pattern and Function Connection* is designed to help students make this transition.

We also know that the *Principles and Standards for School Mathematics* by the National Council of Teachers of Mathematics (NCTM) is filled with great ideas. Now it is up to those of us in the field to identify curriculum that meets these standards. *The Pattern and Function Connection* strongly addresses the *Principles and Standards* by engaging students in appropriate activities. The NCTM *Principles and Standards* Alignment section, included following this introduction, lists the ways this book facilitates the learning objectives of the *Standards*.

We know that time is precious. You have much to teach and only 180 days or so to do the job. Students often need more experience with a new subject than a textbook can offer, yet you can't afford to spend six to eight weeks on a replacement unit that covers only a single topic. Used collectively as a unit, *The Pattern and Function Connection* is designed to take from two to four weeks. You can spend more time with it if you have some spare days and your students want to continue—we offer plenty of extensions—but we have designed the unit to enable closure before your students become bored.

Finally, we believe that mathematics is not a spectator sport—students learn when they are actively engaged. Hence, we utilize "activities," not "lessons," that require students to be active. With these activities, you will be delightfully surprised at how quickly your students will recognize, graph, analyze, and best of all, solve linear functions.

NCTM *Principles and Standards* Alignment

Below are highlights from the NCTM *Principles and Standards for School Mathematics*, Grades 6–8, that are addressed throughout *The Pattern and Function Connection*. The numbers in parentheses refer to the specific activities that help fulfill each standard. Use this alignment in conjunction with the activity objectives stated clearly in the teacher notes to help select appropriate activities for your curriculum.

Number & Operations: Work flexibly with fractions, decimals, and percents to solve problems (3, 5, 8); develop meaning for integers and represent and compare quantities with them (2, 3, 5).

Algebra: Represent, analyze, and generalize a variety of patterns with tables, graphs, words, and, when appropriate, symbolic rules (all activities); relate and compare different forms of representations for a relationship (2, 3, 4, 5, 6, 7, 11); identify functions as linear or nonlinear and contrast their properties from tables, graphs, or equations (2, 6); develop an initial conceptual understanding of different uses of variables (2, 10); explore relationships between symbolic expressions and graphs of lines, paying particular attention to the meaning of intercept and slope (3, 5, 8, 9, 10, 11); use symbolic algebra to represent situations and to solve problems, especially those that involve linear relationships (7, 8, 9, 10, 11); recognize and generate equivalent forms for simple algebraic expressions and solve linear equations (3, 7, 10, 11); model and solve contextualized problems using various representations, such as graphs, tables, and equations (7, 8, 9, 10, 11); use graphs to analyze the nature of changes in quantities in linear relationships (2, 5, 9, 11).

Geometry: Use geometric models to represent and explain numerical and algebraic relationships (1, 2, 4, 5, 6, 11).

Problem Solving: Build new mathematical knowledge through problem solving (all activities); apply and adapt a variety of appropriate strategies to solve problems (all activities).

Reasoning & Proof: Examine patterns and structures to detect regularities (1, 6, 9, 11); formulate generalizations and conjectures about observed regularities (1, 2, 3, 6, 9, 11).

Communication: Organize and consolidate mathematical thinking through communication (all activities); use the language of mathematics to express mathematical ideas precisely (all activities).

Connections: Recognize and use connections among mathematical ideas (2, 3, 4); understand how mathematical ideas interconnect and build on one another to produce a coherent whole (2, 3, 4, 6); recognize and apply mathematics in contexts outside mathematics (7, 8, 9, 11).

Representations: Select, apply, and translate among mathematical representations to solve problems (2, 4, 5, 7, 8, 9, 11).

Teaching *The Pattern and Function Connection*

To help you prepare and organize, teacher notes are provided for each activity and divided into the following sections. Teacher notes will help you integrate the activities as a unit or help you select the best solo activities to supplement your standard curriculum.

Objective

A specific objective for each activity is clearly stated in this section. The objective will help you identify the desired learning outcomes and allow you to correlate the activity to other materials you may be using.

Materials

A teacher's time should not be spent collecting truckloads of materials. We are teachers too, so we've kept the materials simple.

- You will need an overhead projector and one transparency of each of the Pattern Masters found in Appendix A. Activities 3 through 11 also provide transparency masters following the teacher notes for these activities.

- Colored pencils are useful for several of the activities. They allow students to graph three or four functions on one graph and then compare those functions.

- Activity 9 requires one nut and one carriage bolt per student. These are available at hardware stores.

- You will need sufficient copies of the worksheet and homework blackline masters.

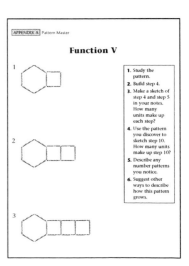

You may also want to make "task cards"—laminated copies of the Pattern Masters to be used by student groups. Alternatively, you could make larger task cards to post for the entire class using construction paper and flat toothpicks or construction-paper polygons. (You may wish to assign creating task cards as an extra credit assignment or have a student aide make them for you.) Although more-experienced students will not have difficulty working from transparencies of the patterns, students working in small groups will appreciate having their own task cards. Attach a copy of the Task Card Instructions, page 88, to each card as shown.

Time Required

This section of the teacher notes suggests a recommended number of class periods that you may want to take for each activity. As a unit, the activities will require about twelve to twenty-two class periods. To that total you may wish to add days for reviewing homework, a test, or other aspects of your existing curriculum (such as problems-of-the-week). You may also select individual activities that address specific goals of your curriculum, or stagger the activities throughout the year to provide your students ongoing practice with functions.

Suggested Teaching Procedures

Suggested Teaching Procedures provide detailed instructions for presenting each activity. Keep in mind that these procedures are merely guidelines—you will find your own improvements and additions. Use the procedures until you gain confidence and expertise at leading your students through each activity.

Journal Prompts

As shown in the Communication strand of the NCTM *Principles and Standards*, the use of verbal language helps students process and organize their mathematical thinking. We also feel that writing is a great tool for assessing what a student truly knows. For these reasons journal prompts are suggested for each activity.

Each journal prompt may be used as a warm-up or introduction to the lesson, or as reflective writing to allow students to synthesize and process the day's learning. Alternatively, you may wish to use these prompts as class discussion starters. Either way, the importance of language in processing mathematical thought should not be overlooked.

We have found in teaching these activities that although many students have excellent skills in thinking algebraically, they are inexperienced in translating their thoughts into symbolic algebra. These students can correctly describe in words the pattern they see and make accurate predictions about its extension, but they don't necessarily know how to write a formula for what they clearly hold in their minds. The more we teach this unit, the more we see algebra as a language itself—it is the language of patterns. Journal prompts can help students make further connections between verbal language and the language of algebra.

Assessment

Assessment guidelines are provided for each activity. These are often informal assessments based on your observations of your students. Although a Linear Functions Test is provided in Appendix C, we know as fellow teachers that there is no better assessment of students than their teachers' observations.

Homework

Homework is provided for each activity to allow students to practice what they have learned. Although the homework is a reflection of the activity, it is practice-based rather than investigative. We have tried to make the classroom the arena for discovery and home the arena for practice. This makes it easier for students, since you will be present during their discovery process when many of their questions arise.

Extensions

Extensions are also suggested for each activity. You may use these to take your students further along in an activity, or to explore new areas. Classes that have had more experience in algebra may use these extensions for homework. Extensions can also be used as the basis for investigations or other problem-solving activities such as problems-of-the-week. The problems found in Appendix B: More Problem-Solving Activities can also be used as extensions to supplement your in-class activities.

1 Exploring Patterns

Objective

Students will study patterns and make predictions about their extensions.

Materials

8 to 10 linear Pattern Masters (task cards or transparencies), such as Functions A, B, C, D, E, F, G, and H

2 to 4 nonlinear Pattern Masters (task cards or transparencies), such as Functions O, P, R, and GG (optional)

Worksheet 1: Exploring Patterns (2 pages)

Homework 1: Exploring Patterns (4 pages)

Pattern blocks

Toothpicks

Time Required

1 or 2 class periods

Journal Prompts

- What pattern was the most interesting to you? Why? Sketch it in your journal.

- How were you able to predict what step 10 would be? Give an example.

- You are beginning a unit on algebra. What is algebra? What people would use algebra in their lives?

Assessment

Check student journals by reading them or having students read them to the class. As students study the patterns, ask them if they could extend the patterns to higher numbers. Check their results to see if they are correct.

Homework

Homework 1: Exploring Patterns
This assignment provides further practice in exploring functions. Students should sketch steps 4 and 5 and make predictions about future steps as they did in class. More formal analysis of these patterns will come in subsequent activities. It will not be necessary to collect this assignment the next day, as the students will be using it to complete Homework 2.

Extensions

Have students design their own toothpick patterns and make predictions about them.

Suggested Teaching Procedure

1. Students should work in groups of three or four for this activity. Hand out Worksheet 1: Exploring Patterns. Distribute pattern blocks and toothpicks. As you model this activity, post a task card or display a transparency of one pattern, such as Function A, where all students can see it.

2. On the worksheet, have students fill out the answer section for the first function based on the pattern you have chosen. Students should first use pattern blocks or toothpicks to build their designs, then they should sketch the fourth and fifth steps with you together as a class. After completing the fifth step, ask students to predict the tenth step. Allow for a variety of student guesses. Verify or reject the guesses through class discussion. You may find that more than one right answer can be verified because some students will see several correct ways to continue the pattern.

3. It is important to promote discussion and encourage students to describe precisely what they see. Students should write their descriptions in the section of the worksheet that says, "What patterns do you notice?" For example, if the steps on the task card correspond to values of 5, 9, and 13, some students may say, "It goes up by four each time." You can then help them to clarify by saying, "Do you mean the pattern goes four, eight, twelve?" When they say that the pattern is one number higher than that, you can have them write, "It starts with one, then goes up by four each time." This is critical to future connections to the expression $4n + 1$ and the formula $f(n) = 4n + 1$.

4. After modeling the activity, pass out one task card or transparency per group and have students work on it. During this time of exploration, circulate around the room. Students will have questions about procedure, how to record their work, and what to do next. Use this time to ask

leading questions such as "How do you know your tenth step is correct?" Students will often try to merely double the fifth step. Show them that this is incorrect by doubling the second step and comparing it to the fourth step.

Check to ensure that students actually build at least the fourth step with pattern blocks or toothpicks. There may be a tendency among some group members to omit this step. Often this hinders the progress of students who need to see this concrete step.

5. For each pattern, students are asked, "Can you suggest other ways this pattern might grow?" With some patterns, students may disagree about how it progresses. This divergent thinking is good and should be encouraged. Although one group may see one pattern, another group or individual may see a different pattern entirely. Ask students to explain their reasoning. If the pattern can be justified, it should be considered valid. For example, in the nonlinear pattern 1, 2, 4, some students will say the terms are doubling and that step 4 will have a value of 8. However, some students may notice that you can add 1 to the first term to get the second term, then add 2 to the second term to get the third term. These students may claim that you must then add 3 to get the next term. This results in step 4 having a value of 7. Both of these patterns could be true based on the information given.

6. After a group of students has completed a single pattern, they may select a new pattern. You may prefer to wait until a majority of the groups have finished and then rotate patterns among the groups. This ensures that all groups work on the same cards, but also has the drawback that some cards require much more time than others.

Extra task cards or transparencies should be available for groups that finish early. It is also advisable to have a few challenging cards for students who are very good at this activity. You may want to include some nonlinear patterns from the back of this book.

7. One class period of work on these patterns may be sufficient if students are able to see the relationships easily. In most cases, two periods are better. If students become restless with the activity, it might be a good time to begin a discussion with the whole class. Ask teams to share their results on a particular function. Ask other teams if they agree or if they can see different patterns.

Answer Key

Homework 1: Exploring Patterns

Step 10:

Function 1:   21
Function 2:   41
Function 3:   32
Function 4:   22
Function 5:   43
Function 6:   61

Steps 4, 5, 6, and 10:

Function 7:   20, 24, 28, . . . , 44
Function 8:   19, 23, 27, . . . , 43
Function 9:   23, 28, 33, . . . , 53
Function 10:   39, 47, 55, . . . , 87
Function 11:   37, 41, 45, . . . , 61
Function 12:   75, 81, 87, . . . , 111
Function 13:   143, 157, 171, . . . , 227
Function 14:   56, 52, 48, . . . , 32 (This is a decreasing pattern.)
Function 15:   15.5, 18, 20.5, . . . , 30.5
Function 16:   16, 32, 64, . . . , 1024 (This is a doubling pattern.)

Exploring Patterns

For each pattern, follow these instructions:

1. Study the pattern your teacher has provided.
2. Build step 4.
3. Make a sketch of step 4 and step 5. How many units make up each step?
4. Use the pattern you discover to sketch step 10. How many units make up step 10?
5. Describe any number patterns you notice.
6. Suggest other ways to describe how this pattern grows.

Function _____

step 4	step 5	step 10
units: _____	units: _____	units: _____

What patterns do you notice? _____

Can you suggest other ways this pattern might grow? _____

_____

Function _____

step 4	step 5	step 10
units: _____	units: _____	units: _____

What patterns do you notice? _____

Can you suggest other ways this pattern might grow? _____

_____

Function _____

step 4	step 5	step 10
units: _____	units: _____	units: _____

What patterns do you notice? _____

Can you suggest other ways this pattern might grow? _____

_____

Function _____

step 4	step 5	step 10
units: _____	units: _____	units: _____

What patterns do you notice? _____

Can you suggest other ways this pattern might grow? _____

_____

Function _____

step 4	step 5	step 10
units: _____	units: _____	units: _____

What patterns do you notice? _____

Can you suggest other ways this pattern might grow? _____

_____

Name _____

Exploring Patterns

For each pattern, sketch steps 4 and 5. Then predict how many units would be in step 10. Describe the pattern.

Function 1. Count the toothpicks.

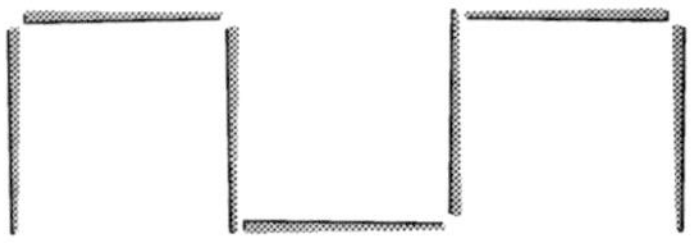

1    2 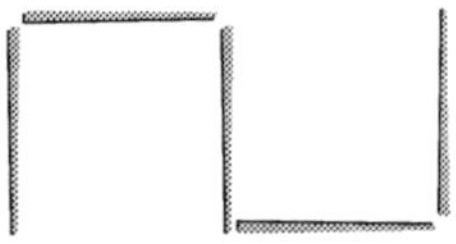   3

4	5

How many would be in step 10? _____
How is the pattern changing or growing? _____

Function 2. Count the toothpicks.

1    2 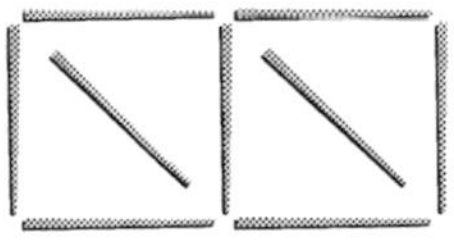   3

4	5

How many would be in step 10? _____
How is the pattern changing or growing? _____

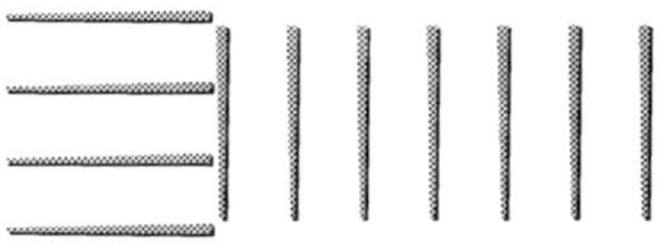

Function 3. Count the toothpicks.

1 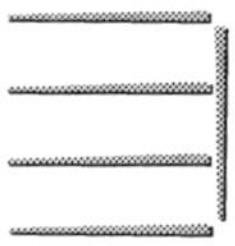   2 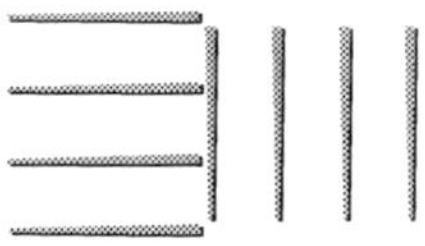   3

4	5

How many would be in step 10? _____
How is the pattern changing or growing? _____

Function 4. Count the toothpicks.

1

2

3

4

5

How many would be in step 10? _____
How is the pattern changing or growing? _____

Function 5. Count the tiles.

1

2

3

4

5

How many would be in step 10? _____
How is the pattern changing or growing? _____

Function 6. Count the tiles.

1

2

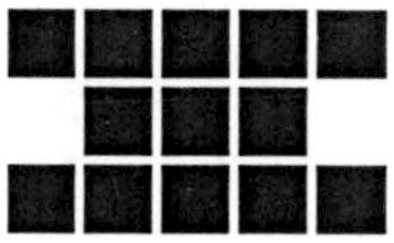

3

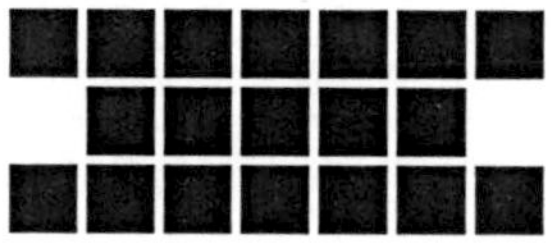

4

5

How many would be in step 10? _____
How is the pattern changing or growing? _____

Find the next three numbers in each pattern. Then find step 10.
Describe in words how the pattern is changing.

Function 7.

step	1	2	3	4	5	6	. . .	10
result	8	12	16	__	__	__		__

Describe the pattern. _____

Function 8.

step	1	2	3	4	5	6	. . .	10
result	7	11	15	__	__	__		__

Describe the pattern. _____

Function 9.

step	1	2	3	4	5	6	. . .	10
result	8	13	18	__	__	__		__

Describe the pattern. _____

Function 10.

step	1	2	3	4	5	6	. . .	10
result	15	23	31	__	__	__		__

Describe the pattern. _____

Function 11.

step	1	2	3	4	5	6	. . .	10
result	25	29	33	__	__	__		__

Describe the pattern. _____

Function 12.

step	1	2	3	4	5	6	. . .	10
result	57	63	69	__	__	__		__

Describe the pattern. _____

Function 13.

step	1	2	3	4	5	6	. . .	10
result	101	115	129	__	__	__		__

Describe the pattern. _____

Function 14.

step	1	2	3	4	5	6	. . .	10
result	68	64	60	__	__	__		__

Describe the pattern. _____

Function 15.

step	1	2	3	4	5	6	. . .	10
result	8	10.5	13	__	__	__		__

Describe the pattern. _____

Function 16.

step	1	2	3	4	5	6	. . .	10
result	2	4	8	__	__	__		__

Describe the pattern. _____

2 Graphing Functions

Note: Activity 1 is a prerequisite for Activity 2. Students will use the previous homework to complete this activity.

Objective

Students will construct T-tables and graphs of functions. They will look for patterns in and similarities between the T-tables and graphs of functions.

Materials

8 to 10 linear Pattern Masters (task cards or transparencies), the same set that you used for Activity 1

2 to 4 nonlinear Pattern Masters (task cards or transparencies), such as Functions O, P, R, and GG (optional)

Worksheet 2: Graphing Functions

Homework 1: Exploring Patterns (retained from Activity 1 with all patterns completed) (1 page)

Homework 2: Graphing Functions (2 pages)

Rulers or straightedges

Colored pencils (optional)

Time Required

1 or 2 class periods

Journal Prompts

- How is a pattern of toothpicks or tiles similar to its T-table? How is it similar to its graph? How is the T-table similar to the graph?

- If your parents asked you what you learned in math class today, what would you tell them?

Assessment

Check students' T-tables and graphs for accuracy as you circulate around the room. Some students will incorrectly count spaces on their graphs or try to make all lines pass through the origin.

Homework

Homework 2: Graphing Functions
In order to complete this assignment, students must have access to their completed Homework 1: Exploring Patterns. It will not be necessary to collect Homework 2 the next day because students will be using it to complete the homework assignment for Activity 3.

Extensions

Have students design a pattern of their own. Then have them construct a T-table and graph. Allow students to share their creations with the class. Which graphs are straight lines? Which aren't straight? Why?

Suggested Teaching Procedure

1. Students should work in groups of two to four for this activity. Hand out Worksheet 2: Graphing Functions. Post a task card of Function A where all students can see it, or use a transparency of this pattern as you model this activity. The students should label the first T-table on the worksheet with the appropriate letter, "A."

2. Prior to graphing ordered pairs, students should label the axes of their graphs. The horizontal axis should have a scale of one. Thus the students should label the axis by counting by ones, starting with zero. The vertical axis should be labeled in a similar fashion. Every unit does not need to be labeled. Some students may find it helpful to use tick marks (small lines to the left of or below the axis) to facilitate counting. An example of using tick marks to label the vertical axis by fives is shown. Check to ensure that students are numbering each grid line, not each space between lines.

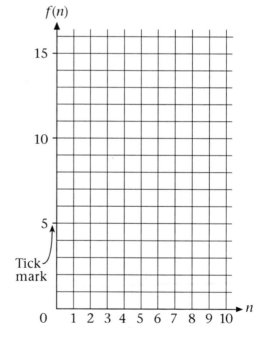

3. Show the students how to plot ordered pairs. To begin, refer to the first step of the pattern on the task card or transparency you have chosen. Ask students to identify the number of tiles or toothpicks used. For

Function A, the first step has 4 tiles. Next, have the students enter (1, 4) in the T-table. You may wish to tell them that the "*n*" stands for "number of the step" and refers to the number of spaces they will count to the right. The "*f(n)*" means "value of the function or pattern at the step" and tells them how many spaces to count upward. Although this is an informal definition, it works well with students who are beginning a study of functions. Lastly, show the students how to plot the ordered pair (1, 4). Repeat this modeling process for steps 2 and 3 of the pattern. Students will soon catch on to the procedure.

4. After students have plotted the first three points of the function, ask them what they notice about the points. They will say that the points are in a straight line. They may also notice that the points are evenly spaced. These observations are critical.

5. Have students complete the T-table for steps 4 and 5 and graph each point in sequence so that they see the interrelationships between table and graph. (Note that there is a space above the number *n* = 1 on the T-table. The number *n* = 0 will be added here in step 7.) Although the fourth and fifth steps of the function are not presented on the task card or transparency, students will have no trouble understanding where they should go graphically—they will visually continue the linear progression of the previous points. You may want to model plotting the fourth and fifth points by counting right and upward from the origin each time. This should lead at least one student to comment, "You don't have to count up thirteen spaces. Every time you go right one space, you go up three more from the last point." Encouraging classroom discussion as you model plotting will help students verbalize the concept of slope that will be introduced in the next activity.

6. After plotting the set of five ordered pairs, have students use a ruler to connect the points with a straight line as shown. Note that picture patterns are discrete functions and have defined values only for whole number steps. Technically, the plotted points should not be connected with a straight line. However, we have found that the line helps beginning learners see the linearity of the function. It also allows them to compare multiple functions plotted on the same graph. If you desire, you can explain that the line is there only to indicate direction and does not imply fractional values for *n*.

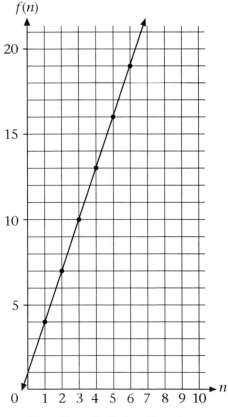

Function A $f(n) = 3n + 1$

7. Insert a 0 on the T-table in the *n* column. Ask what the value of the function would be at step 0. Although many students may initially say it is 0, others will notice that it is 1. Ask both sides to justify their opinion. The former group will probably say that before you begin, you have 0 or nothing. However, other students will note that the function decreases by 3 going backward, and 4 – 3 = 1. While you may want to highlight the point (0, 1) on the graph, it is not necessary to mention the term "intercept" at this point in the unit. Both slope and intercept will be addressed in context in Activity 3.

8. Indicate the increase on your T-tables as shown in this figure. This will help students relate to the concept of slope later on. Again, it is not necessary to introduce the term "slope" at this time. Show that the increase is +3 so that when a negative slope is introduced it can be marked with negative numbers.

n	*f(n)*
0	1
1	4
2	7
3	10

+3
+3
+3

9. Students should be ready to find step 17. There may be different answers—ask how they got each one. The most common wrong answer is 51. They get this from 3 × 17 = 51. This answer can be disproven by trying the method with known values in the T-table: 3 × 1 would have to be 4, 3 × 2 would have to be 7, and 3 × 3 would have to be 10. Students will see that each correct answer is actually one more than predicted. Thus for *n* = 17, *f(n)* = 3 × 17 + 1 = 52.

Another way to illustrate the pattern is by making a T-table in this fashion:

n	*f(n)*	meaning
0	3 × 0 + 1 = 1	three tiles in 0 rows plus one
1	3 × 1 + 1 = 4	three tiles in 1 row plus one
2	3 × 2 + 1 = 7	three tiles in 2 rows plus one
3	3 × 3 + 1 = 10	three tiles in 3 rows plus one
17	3 × 17 + 1 = 52	three tiles in 17 rows plus one
1000	3 × 1000 + 1 = 3001	three tiles in 1000 rows plus one

This illustration leads students more naturally to algebraic notation since for step *n*, there would be three tiles in *n* rows plus one, or 3*n* + 1.

10. Ask students what rule works in general. They will say you need to multiply the number by 3 and add 1. On the board write "Three times the number plus one." Ask students, "Is this what you mean?" When they say yes, tell them the rule is written in algebra as 3 × *n* + 1. You may wish to mention that in algebra the multiplication sign is often omitted

so it isn't confused with the letter *x*. Students may then write "3*n* + 1" as the expression for the *n*th term in the T-table.

Students may write the expression in another form. Research has shown that the form 1 + 3*n* is more natural to the context. You will have to decide if this is acceptable. Obviously, it is not the form used in many formal algebra classes. However, for much younger students, adherence to convention should not be allowed to inhibit the development of conceptual understanding. If young students learn a nonstandard form, this can be corrected easily in an algebra class later.

11. Pass out the remaining task cards or transparencies used in Activity 1. Students should work in groups to complete a T-table and graph for each function. They may use the same graphing grid for three separate functions. Colored pencils are helpful to distinguish the three lines on each grid and allow critical comparisons among functions.

12. As groups begin to work independently, check that they have labeled their axes correctly. Students often try to write the number 1 at the origin instead of writing 0. Others may try to number the spaces on the graph instead of the tick marks. You should also check that they label each function, both the T-table and the line, with the appropriate letter. You will probably see numerous incorrect expressions, but it isn't critical to correct them at this point. In fact, many students will not be able to find algebraic expressions. Let them know that they may leave the expression incomplete if they are unsure. It is better to make allowances than to have incorrect expressions copied from group members. Ask leading questions such as "Have you checked your expression to see if it works at all the known values?"

Students may also notice that the functions that were the most difficult to predict in Activity 1 do not graph as straight lines. These are called "nonlinear" functions and are more complex; some attention will be given to them later. Examples are Functions O, P, R, and GG. Some students interpret Function E as a nonlinear function.

13. For Homework 2, students will make graphs and T-tables of some of the functions from Homework 1. Thus they will need to have finished the previous assignment to be successful at this task. Again, it is unnecessary to collect this assignment the following day. Students will subsequently use Homework 2 to complete the homework for Activity 3. Furthermore, students should be instructed that they need not write expressions for Homework 2; they will also do this when they complete the homework assignment for Activity 3.

Homework 2: Graphing Functions

For Activity 2, Homework 2 should have T-tables and graphs completed with the following coordinates:

Function 1

n	$f(n)$
0	1
1	3
2	5
3	7
4	9
5	11
10	21
17	35

Function 2

n	$f(n)$
0	1
1	5
2	9
3	13
4	17
5	21
10	41
17	69

Function 3

n	$f(n)$
0	2
1	5
2	8
3	11
4	14
5	17
10	32
17	53

Function 4

n	$f(n)$
0	2
1	4
2	6
3	8
4	10
5	12
10	22
17	36

Function 5

n	$f(n)$
0	3
1	7
2	11
3	15
4	19
5	23
10	43
17	71

Function 6

n	$f(n)$
0	1
1	7
2	13
3	19
4	25
5	31
10	61
17	103

Function 11

n	$f(n)$
0	21
1	25
2	29
3	33
4	37
5	41
10	61
17	89

Function 12

n	$f(n)$
0	51
1	57
2	63
3	69
4	75
5	81
10	111
17	153

Function 13

n	$f(n)$
0	87
1	101
2	115
3	129
4	143
5	157
10	227
17	325

Function 14

n	$f(n)$
0	72
1	68
2	64
3	60
4	56
5	52
10	32
17	4

Function 15

n	$f(n)$
0	5.5
1	8
2	10.5
3	13
4	15.5
5	18
10	30.5
17	48

Function 16

n	$f(n)$
0	1
1	2
2	4
3	8
4	16
5	32
10	1,024
17	131,072

Name _____

Graphing Functions

Function _____

n	$f(n)$
1	
2	
3	
4	
5	
⋮	
17	
n	

Function _____

n	$f(n)$
1	
2	
3	
4	
5	
⋮	
17	
n	

Function _____

n	$f(n)$
1	
2	
3	
4	
5	
⋮	
17	
n	

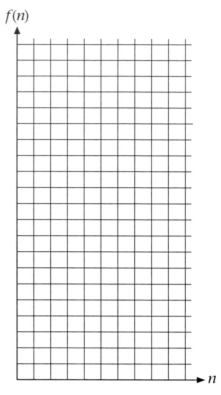

Function _____

n	$f(n)$
1	
2	
3	
4	
5	
⋮	
17	
n	

Function _____

n	$f(n)$
1	
2	
3	
4	
5	
⋮	
17	
n	

Function _____

n	$f(n)$
1	
2	
3	
4	
5	
⋮	
17	
n	

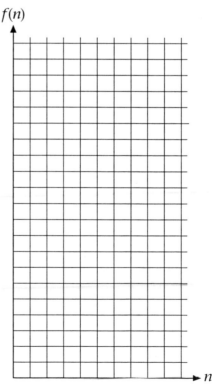

Graphing Functions

Construct a T-table for each of these functions from Homework 1: Exploring Patterns. Then graph each function on the grid. Determine the value of the function at step 0 and put this in the table too.

Function 1

n	$f(n)$
1	
2	
3	
4	
5	
⋮	
10	
⋮	
17	
n	

Function 2

n	$f(n)$
1	
2	
3	
4	
5	
⋮	
10	
⋮	
17	
n	

Function 3

n	$f(n)$
1	
2	
3	
4	
5	
⋮	
10	
⋮	
17	
n	

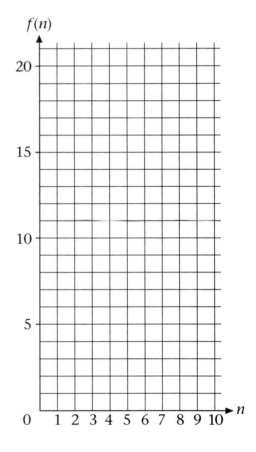

Function 4

n	$f(n)$
1	
2	
3	
4	
5	
⋮	
10	
⋮	
17	
n	

Function 5

n	$f(n)$
1	
2	
3	
4	
5	
⋮	
10	
⋮	
17	
n	

Function 6

n	$f(n)$
1	
2	
3	
4	
5	
⋮	
10	
⋮	
17	
n	

Construct a T-table for each of these functions from Homework 1:
Exploring Patterns. Then graph each function on the grid. Determine
the value of the function at step 0 and put this in the table too.

Function 11

n	$f(n)$
1	
2	
3	
4	
5	
⋮	
10	
⋮	
17	
n	

Function 12

n	$f(n)$
1	
2	
3	
4	
5	
⋮	
10	
⋮	
17	
n	

Function 13

n	$f(n)$
1	
2	
3	
4	
5	
⋮	
10	
⋮	
17	
n	

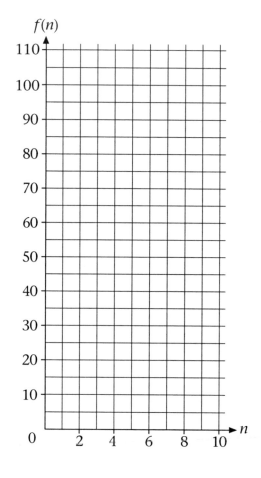

Function 14

n	$f(n)$
1	
2	
3	
4	
5	
⋮	
10	
⋮	
17	
n	

Function 15

n	$f(n)$
1	
2	
3	
4	
5	
⋮	
10	
⋮	
17	
n	

Function 16

n	$f(n)$
1	
2	
3	
4	
5	
⋮	
10	
⋮	
17	
n	

3 Slope and Intercept

> Note: Activities 1 and 2 are prerequisites for Activity 3. The homework assignments progressively build on one another.

Objective
Students will discover patterns in the T-tables and graphs of functions that lead them to an understanding of slope and intercept.

Materials
Pattern Masters (transparencies only) of Functions S, T, U, V, W, X, Y, and Z

Transparency Master 3: Graphing Overlay 1

Clean copies of Worksheet 2: Graphing Functions (see Activity 2) (1 page)

Homework 2: Graphing Functions (retained from Activity 2 with T-tables and graphs complete)

Colored pencils (optional)

Time Required
1 class period

Journal Prompts
- What is "slope"? Give an example. What is "intercept"? Give an example.

- How could you find the slope of a function on a graph? How could you find the slope on a T-table?

- How could you find the intercept of a function on a graph? How could you find the intercept on a T-table?

| **Assessment** | Check students' expressions as you circulate around the room. Ask students to make predictions about the expression that a pattern produces. |

| **Homework** | Complete Homework 2: Graphing Functions |

Have students enter the *n*th term expressions for Functions 1 to 6 and 11 to 15 from Homework 2. They may also try Function 16 if they wish.

| **Extensions** | Once students can find an algebraic expression, have them use it to evaluate the function at higher levels such as *n* = 143. |

Have students make task cards of their own patterns. Then they can exchange them with classmates, construct T-tables, graph them, and find the expressions.

Suggested Teaching Procedure

1. Students may work individually or in small groups for this activitiy. Pass out clean copies of Worksheet 2: Graphing Functions. Have students label the axes as they did in Activity 2. Put the transparency of Function S on the overhead projector. Pass out colored pencils if you are using them.

2. Have the students complete the first T-table and graph the function as you do the same on the overhead projector. The function value at *n* = 0 should also be added to the table and graph. Graphing Overlay 1 can be placed atop the transparency of Function S. You will graph three or four functions (S, T, U, and V) on the same overlay, similar to what the students do on their worksheets.

3. Some students who think symbolically may discover an expression for the *n*th term. Test all suggested expressions to see if they work. Sometimes a student will find a linear formula that works for one value of *n* but not for others. This is because many incorrect lines may pass through a single point, but only one straight line may be drawn between two distinct points. For example, with Function S, the formula $f(n) = 4n + 1$ works when $n = 2$, producing $f(n) = 9$. However, when $n = 3$, $f(n) = 13$ instead of 12. A correct linear formula will work at all values of *n*. The formula $f(n) = 3n + 3$ yields both the values of 9 and 12 respectively and works for all other values of *n*. [You may want to broaden students' understanding of function notation by writing the formula as $S(n) = 3n + 3$.] Enter $3n + 3$ in the table for the *n*th term.

4. When students have finished Function S, display Function T. Place Graphing Overlay 1 atop this transparency; clean only the T-table so that the graph of Function S is still visible. Have students exchange colored pencils so that the next function on their graph is a different color. Repeat teaching procedure steps 2 and 3.

It is at this stage that students will notice similarities and differences between the two graphs. If not, ask leading questions about the two lines. One pair of students put it nicely. After graphing Function T, Becky exclaimed, "They're the same!" Her partner countered, "No, they're not." When asked to explain, Becky said the lines were "going the same way." Her partner answered, "But one is higher than the other."

5. Explain to the students that there is a word for the direction of a line. It is borrowed from snow skiing. The "steepness" of a line is called its "slope." These lines have the same slope.

6. Ask the students to compare the two expressions, $3n + 3$ and $3n + 4$. What numbers do they have in common? Someone will state that both expressions multiply by 3, but the amount they add is different. Show students that when two lines have the same slope, the number by which you multiply is the same. In fact, that number *is* the slope, since it determines how much you increase at each step of the pattern. Someone may also say that the number by which you multiply is the amount by which you increase $f(n)$ as you move 1 unit to the right on the graph. These are redundant statements, but they are the very connections we want students to make as they move through multiple representations of function: manipulative, linguistic, graph, T-table, and expression.

The **Slope** Connection

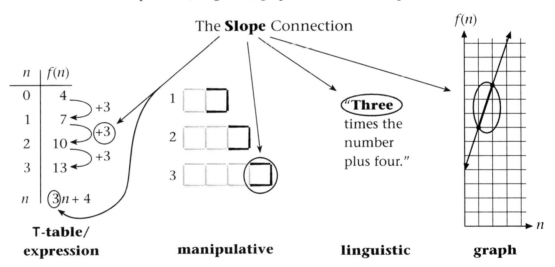

T-table/ expression   **manipulative**   **linguistic**   **graph**

Some students may ask why you need to *multiply* by 3 when the T-table says that you *add* 3 each time. The answer is that if you have to add 3 over and over, you use multiplication; multiplication is "repeated addition."

7. Now it's time to test these discoveries. Put up the transparency for Function U, and repeat teaching procedure steps 2 and 3. Ask the

students if the connections still exist. You will notice a surprising increase in the number of students who can find the expression $3n + 5$.

8. Repeat procedure steps 2 and 3 with Function V, $3n + 6$, if necessary. Using Function V will also require students to create their own T-tables.

9. Now put up the transparency for Function W and have students repeat procedure steps 2 and 3 again. Students should begin using the second half of Worksheet 2; likewise, completely clean Graphing Overlay 1 to start anew. Many students will expect the expression to be $3n + 7$, and they may blindly write it without checking. However, most will see that this expression doesn't work. Although the slope can be determined, that "other number" is not so easy to find. The correct formula is $f(n) = 2n + 1$. Students should enter $2n + 1$ as the expression for the nth term.

10. Next use Function X. Some students will notice that the slope is different. Ask them what remains the same. Students will see that the lines have a point in common at $(0, 1)$ and that this point shows up on both T-tables at $n = 0$. They can also see that in both expressions, you must add 1: $2n + 1$ and $3n + 1$. Tell students this point has a name—it is called the "intercept." For further clarification, have students imagine a quarterback throwing a football down the line of the function, while at the same time, a defensive back runs up the vertical axis and grabs the ball at the point the lines cross. This point is the intercept.

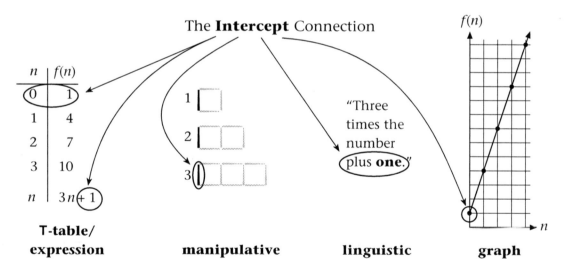

The **Intercept** Connection

T-table/ expression	manipulative	linguistic	graph

Sometimes, students will confuse the slope and intercept. Here is a handy way to distinguish them:

Slope: If the line goes up three each *time*, then that is what you "*times*" by.

Intercept: Referring students' attention to the T-table, tell them this rhyme, "The zero step is the intercept."

---

11. Display Function Y. Students will soon see that the function has the same intercept but that the slope has increased: $4n + 1$. Most of the class will now be writing expressions and formulas easily.

12. Repeat the process for Function Z, $5n + 1$, if necessary. You may also wish to let students explore other patterns if time allows. Consult the Pattern Masters Key in Appendix A for linear examples.

13. Show students how a formula allows us to immediately evaluate a function for any value of n. For example, when $n = 143$, the value of Function Z is $5 \times 143 + 1 = 716$. The one-millionth term is 5,000,001!

14. This will be a successful and encouraging activity for many students making their first expedition into the world of algebra. You will soon see that students can move from one format to the other with ease. From a pattern they can generate a T-table, graph, and formula. They can also generate a graph and formula when given a T-table, a T-table and formula when given a graph, or a T-table and graph when given a formula.

15. Ask the students to translate this sentence into an expression or a formula: "The pattern starts with five, then goes up seven each time." An appropriate expression is $7n + 5$. Next, ask them to write a sentence for this formula: $f(n) = 12n + 32$. One reply may be "The pattern starts with thirty-two, then goes up twelve each time."

16. Ask the students to find expressions for the functions on Homework 2 based on the T-tables and graphs they previously completed. Function 16 may be challenging since it is nonlinear.

After this activity, collect Homework 1 and Homework 2. Check the nth term expressions on Homework 2 for accuracy. If they are correct, then the prior work should also be correct.

Answer Key

Homework 2: Graphing Functions

For Activity 3, the T-tables on Homework 2 should be completed with the following expressions:

Function 1:	$2n + 1$		Function 11:	$4n + 21$
Function 2:	$4n + 1$		Function 12:	$6n + 51$
Function 3:	$3n + 2$		Function 13:	$14n + 87$
Function 4:	$2n + 2$		Function 14:	$-4n + 72$
Function 5:	$4n + 3$		Function 15:	$2.5n + 5.5$
Function 6:	$6n + 1$		Function 16:	2^n

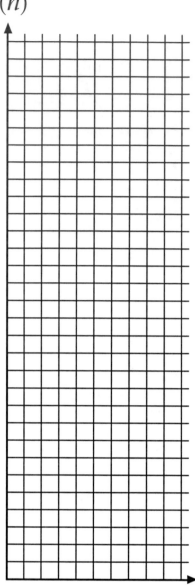

n	$f(n)$
0	
1	
2	
3	
⋮	
17	
n	

4 Working Backward

Objective	Students will solidify understanding of the relationship between a pictorial representation and its formula by creating the pictorial model from the algebraic representation.

Materials	Transparency Master 4: Working Backward
	Worksheet 4: Working Backward (1 page)
	Homework 4: Working Backward (1 page)
	Pattern blocks (optional)
	Toothpicks (optional)

Time Required	1 class period

Journal Prompts	• Does a formula always generate only one pictorial representation? Why or why not? Does a formula always have only one graphical representation? Why or why not?
	• A student described her function this way: "It increases by five each time, beginning with two." Write a formula for this function, and design a pictorial representation of it.

Assessment	Have students exchange their pictorial representations and see if classmates can generate the desired functions.

Homework 4: Working Backward

Students can write their own formulas and design pictorial representations for them. The patterns can be linear or nonlinear. Students can trade with classmates who then try to find the corresponding functions.

You may also wish to give students further practice in graphing functions during this activity.

Suggested Teaching Procedure

1. Students may work in small groups or individually on this activity. Display Transparency Master 4: Working Backward and cover the pictorial representations so that only the algebraic formulas are visible. Direct the students' attention to the function labeled a. Ask them how its pictorial representation might look. You may wish to have some students draw possibilities on the board.

2. Now reveal the full transparency. Have students try to match the two pictures to two of the functions. You may need to encourage them to think about the slope and intercept represented in the function and how this translates into a picture pattern. (The correct pairings are 1/c and 2/a.) Ask them to volunteer their results. Ask if there are other ways to represent the function labeled c besides the one on the transparency. Have some students share possibilities.

3. Have students draw or build representations for the function labeled b. Ask them to compare their representations with those of other members of their group or class. Have students share some of the possibilities on the board.

4. As an optional activity, you may wish to give each group of students the following four functions.

   1. $f(n) = 3n + 4$
   2. $f(n) = 3n + 1$
   3. $f(n) = 3n - 1$
   4. $f(n) = 4n + 3$

   Ask each member to design a pictorial representation of one function. Once they finish, have students regroup according to similar functions. Thus all of those who represented Function 1 would be in one group, all of those who represented Function 2 in a second group, and so on.

Ask students to compare the different pictorial representations of the given function. They should consider these questions:

Is it easier to see the function in some of the pictures than in others? Why or why not?

Which representations allow you to recognize the slope easily?

Which representations allow you to recognize the intercept easily?

5. Pass out Worksheet 4 and ask the students to complete it individually or in groups.

6. Assign Homework 4: Working Backward.

Answer Key

Worksheet 4: Working Backward

1. g
2. c
3. j
4. a
5. h
6. d

Here are three functions and two pictures.
Match the pictures with two of the functions.
Design a picture to illustrate the other function.

a. $f(n) = 2n + 4$ **b.** $f(n) = 4n + 2$ **c.** $f(n) = 4n - 2$

1. 1

2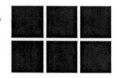

3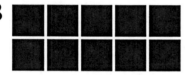

2. 1

2

3

   The Pattern and Function Connection / Copyright © 2001 Key Curriculum Press

Name _____

Working Backward

Below are ten functions and six pictures. Below each pictorial representation, write the letter of the matching function. On the back of this sheet, design a pictorial representation for the four functions that are not pictured. Label them with their functions.

a. $f(n) = 4n - 3$
b. $f(n) = 3n + 2$
c. $f(n) = 2n + 2$
d. $f(n) = 4n + 1$

e. $f(n) = 2n + 3$
f. $f(n) = 4n + 4$
g. $f(n) = 3n + 6$

h. $f(n) = 6n - 5$
i. $f(n) = 6n - 1$
j. $f(n) = 6n + 2$

1. count squares

1
2
3

Function _____

2. count circles

1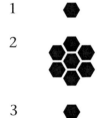
2
3

Function _____

3. count toothpicks

1
2
3

Function _____

4. count squares

1
2
3

Function _____

5. count hexagons

1
2
3

Function _____

6. count squares

1
2
3

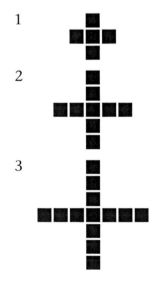

Function _____

Name _____

Working Backward

Below are nine functions. For each one, design a pictorial representation and show three steps. Use polygons, lines, toothpicks, dots, or other shapes. Be creative.

1. $f(n) = 2n + 1$	**2.** $f(n) = 2n$	**3.** $f(n) = 4n + 2$
4. $f(n) = 4n$	**5.** $f(n) = 4n + 4$	**6.** $f(n) = 5n + 3$
7. $f(n) = 2n + 3$	**8.** $f(n) = 3n + 2$	**9.** $f(n) = 2n - 1$

5 Advanced Functions

Objective	Students will develop an understanding of fractional and negative slopes, and negative intercepts.

Materials	Pattern Masters (transparencies only) of Functions KK through NN (negative intercepts), OO through RR (negative slopes), and SS through ZZ (fractional slopes)
	Transparency Master 5: Graphing Overlay 2
	Worksheet 5: Advanced Functions (at least one copy per student is required for each of the three parts of this activity) (1 page)
	Homework 5: Advanced Functions (3 pages)
	Colored pencils (optional)

Time Required	1 to 3 class periods

Journal Prompts	How can you look at a pictorial representation of a function and see that it has

- a negative slope?
- a fractional slope?
- a negative intercept?

Assessment

Check students' expressions and graphs as you circulate around the room. Allowing students to work in small groups helps them to assess their own success by comparing graphs and discussing formulas with peers.

Homework

Homework 5: Advanced Functions
This assignment is easily divided into three sections corresponding to the three parts of this activity. The first four homework functions (A through D) involve negative intercepts; the next four functions (E through H) involve negative slopes; the final four functions (I through L) involve fractional slopes.

Extensions

Fractional slopes can be modeled with other Pattern Masters. For example, with Function A, if squares sell for $0.47, what is the price of the pattern at step 23? This also provides students an opportunity to practice computation with decimals. Alternatively, students can design their own functions involving negative and fractional slopes, and negative intercepts.

Suggested Teaching Procedure

Part 1: Negative Intercepts

1. Students may work individually or in small groups for each part of this activity. Distribute one copy of Worksheet 5 to each student or group.

2. Display a transparency of Function KK. Ask the students to graph steps 1, 2, and 3 as they complete a T-table for those steps. Use Graphing Overlay 2 atop the transparency to simultaneously model the process.

3. Ask students what step 0 would be in the T-table. Some will likely say that step 0 will be 0, since you can't build a picture of negative polygons. While this is true—all polygons have positive areas—the question is referring to the T-table, not the picture. The function steps indicate a slope of 3. Thus, to make the T-table valid, the intercept would have to be −2 since −2 + 3 = 1.

4. Have the students construct a line through the points on the graph. They will see that the line contains the point (0, −2), which is the intercept. Explain that this is called a negative intercept for obvious reasons.

5. Let the students work with Functions LL, MM, and NN. They should construct a T-table and a graph for each function, and find the algebraic expression for the nth term.

Part 2: Negative Slopes

1. Distribute clean copies of Worksheet 5 to each student group.

2. Repeat the procedure from Part 1 using Function OO. Students will see that this function *decreases* by 3 each time. This can be indicated in the T-table as shown.

3. Explain that this is called a negative slope. Function OO has a slope of –3.

4. Students should also explore Functions PP, QQ, and RR.

n	$f(n)$
0	10
1	7
2	4
3	1

(–3 between each)

Part 3: Fractional Slopes

1. Follow the procedure from Parts 1 and 2 using Function SS. As the students fill in the T-table and construct the graph, they will see that the function increases by 3 half circles each time. While most beginning students will say its slope is 1.5 or $1\frac{1}{2}$, you may wish to suggest that they use the more conventional form of $\frac{3}{2}$. (This convention is aligned to the $\frac{rise}{run}$ definition used by many algebra courses.) You could then show them that there is an increase of 3 circles in two steps.

2. Students should then work on Functions TT through XX. Functions UU, VV, and XX involve decimals.

3. Functions YY and ZZ may be used as an extension. These do not give all data points. Students must analyze the given data to determine the missing values and find the intercepts. Function YY should have a slope of $\frac{3}{2}$; Function ZZ, a slope of $\frac{4}{3}$.

Other Options

1. Some of the functions preceding Function KK also involve negative intercepts, negative slopes, or fractional slopes. Students may wish to explore some of them, such as Functions D, G, K, M, and N.

2. Some Pattern Masters use both black and shaded shapes. You may wish to have students explore each color separately. For example, in Function MM, the expression for the black squares is $4n$ and the expression for the shaded squares is $5n - 3$. Combining these two expressions gives us the formula for the total function:

$$
\begin{array}{r}
4n \\
+ \quad 5n - 3 \\
\hline
f(n) = 9n - 3
\end{array}
$$

Although it stands to reason that the sum of the black and shaded squares is the total number of squares, it often surprises students that the sum of their expressions is the expression for the total also. This can be further illustrated with Functions M, R, CC, FF, GG, HH, NN, and QQ.

> Note: Homework 5 may be assigned as a whole after completing Parts 1, 2, and 3, or it may be divided into three corresponding assignments (Functions 1 through 4 for negative intercepts, Functions 5 through 8 for negative slopes, and Functions 9 through 12 for fractional slopes). The latter option may be helpful if you address only one part or complete the activity over several days.

Answer Key

Homework 5: Advanced Functions

Function 1: $3n - 1$

Function 2: $4n - 3$

Function 3: $4n - 2$

Function 4: $6n - 4$

Function 5: $-2n + 10$

Function 6: $-5n + 25$

Function 7: $-2n + 16$

Function 8: $-3n + 19$

Function 9: $\frac{5}{2}n + \frac{1}{2}$

Function 10: $\frac{9}{2}n + \frac{5}{2}$

Function 11: $2.3n + 3.4$

Function 12: $0.45n + 5.45$

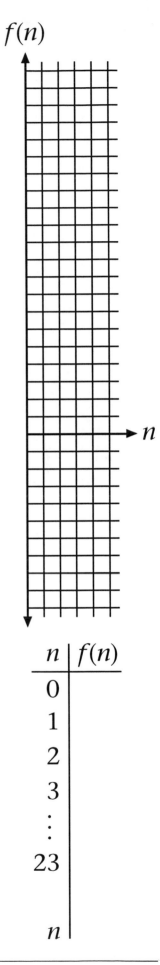

$f(n)$

n	$f(n)$
0	
1	
2	
3	
⋮	
23	
n	

Name _____

Advanced Functions

Function _____

n	$f(n)$
0	
1	
2	
3	
⋮	
23	
n	

Function _____

n	$f(n)$
0	
1	
2	
3	
⋮	
23	
n	

Function _____

n	$f(n)$
0	
1	
2	
3	
⋮	
23	
n	

Function _____

n	$f(n)$
0	
1	
2	
3	
⋮	
23	
n	

Function _____

n	$f(n)$
0	
1	
2	
3	
⋮	
23	
n	

Function _____

n	$f(n)$
0	
1	
2	
3	
⋮	
23	
n	

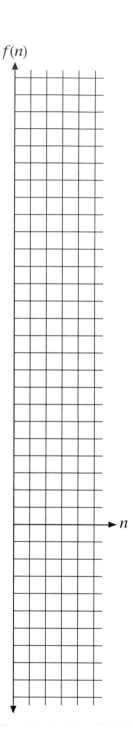

Name _____

Advanced Functions

Fill in a T-table for each function below. Then graph the function.
Write the function's expression in the bottom of the table.

Function 1

1

2

3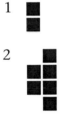

n	f(n)
0	
1	
2	
3	
⋮	
37	
n	

Function 2

1

2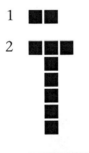

3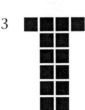

n	f(n)
0	
1	
2	
3	
⋮	
37	
n	

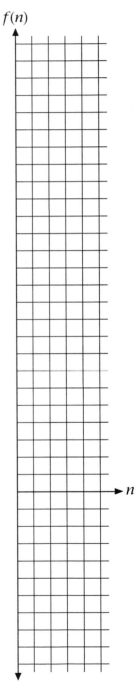

Function 3

1

2

3

n	f(n)
0	
1	
2	
3	
⋮	
37	
n	

Function 4

1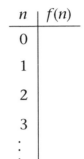

2

3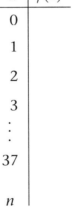

n	f(n)
0	
1	
2	
3	
⋮	
37	
n	

Fill in a T-table for each function below. Then graph the function.
Write the function's expression in the bottom of the table.

Function 5

1

2

3

n	$f(n)$
0	
1	
2	
3	
⋮	
37	
n	

Function 6

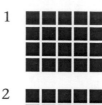
1

2

3

n	$f(n)$
0	
1	
2	
3	
⋮	
37	
n	

Function 7

1

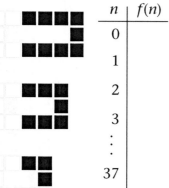
2

3

n	$f(n)$
0	
1	
2	
3	
⋮	
37	
n	

Function 8

1

2

3

n	$f(n)$
0	
1	
2	
3	
⋮	
37	
n	

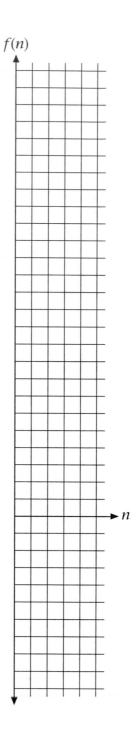

$f(n)$

n

Fill in a T-table for each function below. Then graph the function. Write the function's expression in the bottom of the table.

$f(n)$

Function 9

1

2

3

n	$f(n)$
0	
1	
2	
3	
⋮	
37	
n	

Function 10

1

2

3

n	$f(n)$
0	
1	
2	
3	
⋮	
37	
n	

Function 11

1 5.7

2 8

3 10.3

n	$f(n)$
0	
1	
2	
3	
⋮	
37	
n	

Function 12

1

2 6.35

3

4 7.25

5 7.7

n	$f(n)$
0	
1	
2	
3	
⋮	
37	
n	

$\rightarrow n$

6 Nonlinear Functions

> Note: Some students may find this activity extremely challenging. Analyze your class's level of comfort beforehand. If desired, you can return to this activity as an extension at the end of your unit.

Objective

Students will explore, graph, and construct T-tables for nonlinear functions.

Materials

Nonlinear Pattern Masters (task cards or transparencies), such as Functions O, P, R, AA, BB, CC, DD, GG, HH, II, and JJ

Graph paper (or dot paper)

Rulers or straightedges

Colored pencils (optional)

Time Required

1 to 3 class periods

Journal Prompts

• What makes a function linear or nonlinear?

• Why is it more difficult to find formulas for nonlinear functions?

Assessment

Check student graphs for accuracy. They are a representation of values in the T-table, so if the graph is correct, the T-table is usually correct also.

None.

Students can design and explore their own nonlinear functions.

Suggested Teaching Procedure

Many of the functions in this book are nonlinear. In one sense, that simply means that the graphs are not straight lines. In a more technical sense, their slope is not constant. It varies from step to step, causing the graph to curve. These nonlinear functions are included for two reasons. First, they provide a point of comparison so students can see that a linear function is one that graphs as a straight line. Second, they offer opportunities to explore more complex functions. They show students that not all mathematics is as simple and well behaved as linear functions. Some students will be able to find formulas for these functions. However, they may not use the same methods as they did for finding the formulas of linear functions.

1. Students may work in small groups or individually. Pass out graph paper. (Some students may prefer to use dot paper for graphing.) Instruct students to create a set of axes along the bottom and left sides of the grid. Have them label the axes with scales of 1. This may be the first time students have had to create their own axes on blank graph paper, so circulate to monitor their work.

2. Post a task card or display a transparency of Function HH. Ask the students to construct a T-table for the function and graph steps 0 through 3. T-tables should be constructed on a separate sheet of paper (some students may prefer to use graph paper for this too). Students will notice the points on the graph do not form a straight line. Ask them what occurred in the T-table or in the pattern to cause this. Some may notice that the slope is not constant. (In fact, it increases by the odd numbers. Looking at the differences of the differences shows that the slope has a slope of 2. Since two levels were needed to arrive at a constant difference, this is a second-degree equation.)

3. Ask students how many squares would be in steps 4, 5, and 10. Have them explain how they determined this. Some may say they noticed the pattern of odd numbers. Others may say that step 1 has $1^2 = 1$ square; step 2 has $2^2 = 4$ squares; step 3 has $3^2 = 9$ squares. Thus $4^2 = 16$ squares. This leads to the formula $f(n) = n^2$.

n	$f(n)$
0	0
1	1
2	4
3	9

+3 +2
+3 +2
+3

4. Explore other nonlinear functions. Not all of them can be understood in the same way as Function HH. Those that require second-degree equations are Functions R, AA, BB, CC, GG, II, and JJ. Functions P and DD have alternating slopes and can be described by the greatest integer function. Function O is exponential.

7 Banking on Algebra

Objective	Students will use T-tables and graphs to solve problems involving savings accounts.
Materials	Transparency Master 7: Savings Account Problems $\frac{1}{4}$-inch graph paper (or dot paper) Homework 7: Banking on Algebra (3 pages)
Time Required	1 class period
Journal Prompts	• How is a savings account like an algebra formula? • This formula represents a savings account. What does it tell you about the account? $$f(m) = 25m + 80$$
Assessment	Check student graphs and T-tables for accuracy.
Homework	Homework 7: Banking on Algebra
Extensions	Have students design and solve savings account problems of their own.

ACTIVITY 7   Banking on

Suggested Teaching Procedure

1. Students may work individually or in small groups. Display Transparency Master 7: Savings Account Problems, showing only the first two sentences of problem 1. Students will likely volunteer that they need more information to figure out the problem. Acting confused, ask them what information they need. When they tell you they need to know how many months are involved, answer that you don't know—Randy didn't tell you. Then ask students if they could solve the problem if they knew Randy would make only one monthly deposit. Students may tell you the answer is $40 + $70 = $110. If they do, ask them what would happen if he made two monthly deposits. When students see that the equation for answering this question is $2 \times \$40 + \$70 = \$150$, ask them how much Randy would have if he made zero monthly deposits. The answer to this is $70, the intercept. Then ask students how much Randy's balance goes up each month. They will see it goes up by $40. This is the slope. Ask them if they can now make a T-table showing the information. Most students will see that they can.

2. Pass out $\frac{1}{4}$-inch graph paper to each student. (Some students may prefer to use dot paper for graphing.) They will also need paper on which to make a T-table, but they may use the back or margin of the graph paper. Students will need to create axes and figure out the best way to scale the axes on their graphs to make all the numbers fit. You may wish to tell students this or let them discover the difficulty on their own. Counting by $10 or $20 per line on the vertical axis will work. Counting by two's on the horizontal axis is sufficient; the answers to subsequent questions will not fit on the graph if the horizontal scale is 1 unit. Have students construct a T-table for Randy's account, graph the function, and write the formula: $f(m) = 40m + 70$. Tell them m stands for the number of months. [Students are likely to have used n, so you can now show them that any variable will work. You may want to further extend the concept of variable by using function names other than $f(m)$, such as $g(m)$ or $h(m)$.]

3. Now give some values for m (months). What if Randy makes 17 monthly deposits? What if he makes 37? What if he makes monthly deposits for 4 years (48 months)? Students can now see that even when they don't have enough information to solve a problem, they can find a formula that will solve all related problems. Reveal the remainder of problem 1 on the transparency and have students calculate the answers. Alternatively, students can extend their graphs to answer each query. Respectively, the answers are $270, $470, and $790.

4. Reveal the entirety of problem 2 on the transparency. Most students will construct a second T-table beside the first. Then they will extend the tables until they match. You may also wish to show students that

graphing the two functions shows their intersection as the solution. Lastly, you may wish to show students a more formal algebraic solution:

$$
\begin{aligned}
40m + 70 &= 35m + 310 \\
-35m &\quad -35m \\
5m + 70 &= 310 \\
-70 &\quad -70 \\
\frac{5m}{5} &= \frac{240}{5} \\
m &= 48
\end{aligned}
$$

By substituting 48 months into either equation, students can find that Randy and Angela will each have $1,990.

5. In problems 3 and 4 of the transparency, students are given the value of a savings account after a period of time and are asked to find the period of time. Algebraically, they are given $f(m)$ and asked to solve for m. Many students will rely on a guess-and-check method. Others may extend their T-tables until they reach the desired amount while some may use more formal methods, which you may wish to explore at this time.

For problem 3

$$
\begin{aligned}
40m + 70 &= 1000 \\
-70 &\quad -70 \\
\frac{40m}{40} &= \frac{930}{40} \\
m &= 23.25
\end{aligned}
$$

Since Randy makes deposits monthly, and since 23 months yields only $990, we see that Randy must save for 24 months to have at least $1,000.

For problem 4

$$
\begin{aligned}
35m + 310 &= 1220 \\
-310 &\quad -310 \\
\frac{35m}{35} &= \frac{910}{35} \\
m &= 26
\end{aligned}
$$

Angela will have exactly $1,220 after exactly 26 months.

6. Ask the students to translate this sentence into a formula: "Tomekia had $29 in the bank. She added $22 more each month." They should get $f(m) = 22m + 29$. Next, ask them to write a sentence for this formula: $f(m) = 24m + 112$. One possible answer is "Juan had $112 in the bank. He added $24 more each month."

7. Assign Homework 7. The second page provides more challenging problems. In fact, you may use the last problem to introduce or review the concept of a negative slope.

Answer Key

Homework 7: Banking on Algebra

1. $f(m) = 25m + 75$
2. $f(m) = 14m + 306$
3. After 65 months, Derek will have exactly $1,700.
4. After 21 months.
5. Both Rosa and Derek will have $600.
6. $f(m) = 27m + 72$; after 9 months, San will have enough to buy the bike.
7. $f(m) = 16m + 115$; no, Keisha cannot buy her bike before San; Keisha would need 12 months to have enough money.
8. $f(m) = 32m + 39$; $f(m) = -10m + 435$; after 10 months, Chris can afford the bike.

1. Randy already has $70 in his savings account. He will add $40 per month. How much money will he have?

How much money will he have after the first five months?

How much money will he have after the tenth month?

How much money will he have after the eighteenth month?

2. Angela already has $310 in her savings account. She will add $35 per month. How many months will it take before she and Randy have the same amount of money in their accounts? How much will each of them have?

3. Randy wants to know when he will have at least $1,000. Find a way to determine this for him.

4. Angela finds she has exactly $1,220. How many months has she been saving?

Name _____

Banking on Algebra

1. Derek has $75 in his savings account. He will add $25 every month. Use this T-table to show how Derek's account will grow. Write an expression or formula and make a graph below.

Derek's Account

month	principal
0	
1	
2	
3	
4	
5	
m	

2. Rosa has $306 in her savings account. She will add $14 every month. Use this T-table to show how Rosa's account will grow. Write an expression or formula and make a graph below.

Rosa's Account

month	principal
0	
1	
2	
3	
4	
5	
m	

3. When will Derek have $1,700? _____

4. When will Rosa and Derek have the same amount of money? _____

5. How much money will they each have then? _____

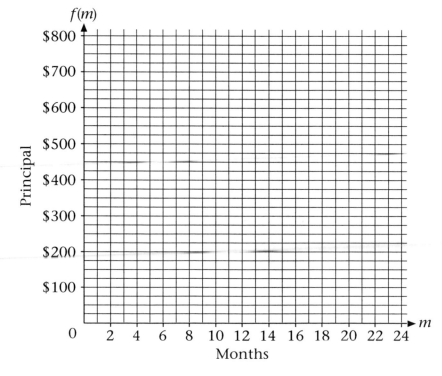

What scale has been used for the horizontal axis? One unit = _____.

What scale has been used for the vertical axis? One unit = _____.

The Pattern and Function Connection / Copyright © 2001 Key Curriculum Press

6. San wants to buy a bicycle for $295. He currently has $72 in his account. If he adds $27 each month, when will he have enough money for the bike? _____

San's Account

month	principal
0	
1	
2	
3	
4	
5	
m	

7. Keisha wants the same bike that San wants. If she already has $115 in her account, and she adds $16 per month, can she buy her bike before San can buy his? _____

Keisha's Account

month	principal
0	
1	
2	
3	
4	
5	
m	

Graph the function for problems 6 and 7 on this grid. Choose a scale that allows the data to fit. Label each function on the graph with its formula.

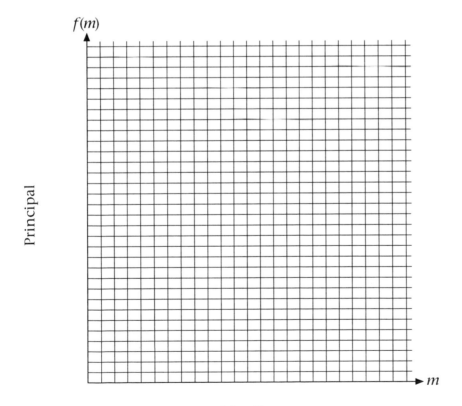

f(m)

Principal

Months

What scale did you use for the horizontal axis? One unit = _____.

What scale did you use for the vertical axis? One unit = _____.

8. Chris has $39 in the bank. He adds $32 to it each month. He wants to buy a bike that costs $435. Every month the bike's price is reduced by $10. How many months will it take for Chris to afford the bike? _____

Chris's Account			Bike Sale	
month	principal		month	bike price
0			0	$435
1			1	$425
2			2	
3			3	
4			4	
5			5	
m			m	

Graph the function for problem 8 on this grid. Choose a scale that allows the data to fit. Label each function on the graph with its formula.

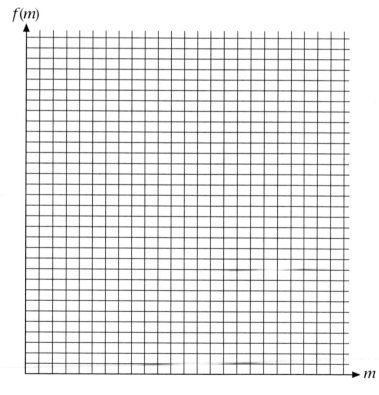

$f(m)$

Principal

Months

m

What scale did you use for the horizontal axis? One unit = _____.

What scale did you use for the vertical axis? One unit = _____.

8 The Long-Distance Connection

Objective	Students will use T-tables and graphs to solve problems involving long-distance phone rates.
Materials	Transparency Master 8: Schedule of Rates Graph paper (or dot paper) Colored pencils (optional) Homework 8: The Long-Distance Connection (2 pages)
Time Required	1 class period
Journal Prompts	• If you saw three lines on a graph representing calls to Sacramento, how could you tell which line was the day rate? Explain your reasoning. • Here are three phone-rate formulas. Label them "day," "evening," and "night." Explain your reasoning. $f(m) = 0.13m + 0.10 \qquad f(m) = 0.15m + 0.11 \qquad f(m) = 0.10m + 0.08$
Assessment	Check student graphs and T-tables for accuracy.
Homework	Homework 8: The Long-Distance Connection

You can vary T-tables to increase difficulty. For example, if values are given for the fourth and seventh minutes only, students must find the slope (additional minute charge) and intercept (connection fee). You could give the nine-minute rate and the connection fee, or the slope and the ten-minute rate.

Actual per-minute charges are not measured in whole cents or whole minutes. Ten-second pricing increments are common as are per-minute rates such as $0.089. These values will make functions more complex.

You can also ask students to research and compare long-distance rates for various companies (MCI, AT&T, Sprint, etc.). Students could also examine differences between home-phone service and cellular service.

Suggested Teaching Procedure

1. Students may work individually or in groups for this activity. Talk to the students about our dependence on phone service. Ask them to describe different ways we use telecommunication (phones, cellular phones, pagers, faxes, phone cards, etc.). You may want to ask students what long-distance calling plans they have at home and how many long-distance calls they make per month.

2. Display Transparency Master 8: Schedule of Rates. Point out to students that rates for using the phone differ depending on the time of day or day of the week that a call is made. Some students may not be aware of this real-world convention. Direct their attention to the cost of a call from Redding to Alturas on a weekday. Make sure students first look at the part of the chart that is shaded black. Now have them identify the two columns of information labeled "1st min." and "add'l. min.," which tell the day rates for a call.

3. Point out that the cost for a weekday call from Redding to Alturas is $0.11 for the first minute. This is true whether you talk for 1 second, 10 seconds, or 60 seconds. But if you talk for 61 seconds (1 minute and 1 second), you have to pay for an entire additional minute at $0.07. You may wish to define this as a "step function," since the cost steps up each minute.

4. Have the students use their own paper and construct a T-table for the day rate from Redding to Alturas. Their table should show "minutes," or m, and cost, or $f(m)$. Ask them what the value is at $m = 0$. Some will say it is zero—if you don't call, you don't pay. However, the phone company does charge more for the first minute, and that translates into the intercept: $0.04. This can be thought of as the cost of connecting. That is, the phone company charges an additional $0.04 to connect each call,

regardless of length. (If the students want to know why the first minute is more expensive, or if they want to know why rates vary according to the time of day, have them call the operator and ask. The phone company has been kind enough to share information with students when they explain that they are studying the algebra involved. Students may also wish to ask why night rates apply all day on weekends.)

5. Students should now be able to write the formula $f(m) = 0.07m + 0.04$. Give students blank graph paper and ask them to graph the function using appropriate scales (a horizontal scale of 1 and a vertical scale of 0.05 is appropriate for the situation). Ask them to identify the slope and the intercept. As mentioned in Activity 2 for picture patterns, this graph is theoretically a collection of discrete points representing whole minute increments; drawing a linear function simply helps students visualize the direction of the graph. For older students, you may want to demonstrate how to properly graph a step function.

6. Have students repeat this procedure for the evening rate from Redding to Alturas. Show them that they need to use the next set of values ($0.09 and $0.06) on the chart. The formula is $f(m) = 0.06m + 0.03$. Ask students what they notice about the graph. They will see it has a lower slope and a smaller intercept. [This activity is another valuable place to introduce varied function notation. For example, use $f(m)$ for day, $g(m)$ for evening, and $h(m)$ for night.]

7. Lastly, repeat the process for the night rate. The formula is $f(m) = 0.04m + 0.03$.

8. Ask students to calculate the cost for a 20-minute phone call from Redding to Alturas at various times of day. The differences may surprise them.

Day	$0.07 \times 20 + $0.04 = $1.44
Evening	$0.06 \times 20 + $0.03 = $1.23
Night	$0.04 \times 20 + $0.03 = $0.83

You may want to show students how to approximate these values on the graphs by locating 20 on the horizontal axis, moving up to the graph of each function, and determining the corresponding cost on the vertical axis.

9. Next, ask students how long they could talk to a friend in Alturas for $1.00 if they called at various times of day.

Day	$0.07 \times 13 + $0.04 < $1.00
Evening	$0.06 \times 16 + $0.03 < $1.00
Night	$0.04 \times 23 + $0.03 < $1.00

Again, they can estimate these on the graphs by locating $1.00 on the vertical axis and determining the corresponding minutes on the horizontal axis.

10. Ask students to translate this sentence into a formula: "A call costs $0.16 for the first minute and $0.10 for each additional minute." They should get $f(m) = 0.10m + 0.06$. Then ask them to write a sentence for this formula: $f(m) = 0.14m + 0.08$. A possible answer is "A call costs $0.22 for the first minute and $0.14 for each additional minute."

11. Assign Homework 8: The Long-Distance Connection.

Answer Key

Homework 8: The Long-Distance Connection

1. Redding to Oroville:
Day	$f(m) = 0.12m + 0.02$
Evening	$f(m) = 0.10m + 0.02$
Night	$f(m) = 0.07m + 0.02$

2. Redding to Red Bluff:
Day	$f(m) = 0.13m + 0.02$
Evening	$f(m) = 0.11m + 0.02$
Night	$f(m) = 0.07m + 0.03$

3. Redding to Challenge:
Day	$f(m) = 0.10m + 0.02$
Evening	$f(m) = 0.07m + 0.03$
Night	$f(m) = 0.05m + 0.03$

Schedule of Rates

	M	Tu	W	Th	F	Sa	Su
8 AM–5 PM	■	■	■	■	■	☐	☐
5 PM–11 PM	▥	▥	▥	▥	▥	☐	☐
11 PM–8 AM	☐	☐	☐	☐	☐	☐	☐

	■ Day		▥ Evening		☐ Night	
From: Redding	1st min.	add'l. min.	1st min.	add'l. min.	1st min.	add'l. min.
To:						
Alturas	$0.11	$0.07	$0.09	$0.06	$0.07	$0.04
Challenge	0.12	0.10	0.10	0.07	0.08	0.05
Millville	0.13	0.11	0.11	0.09	0.08	0.06
Oroville	0.14	0.12	0.13	0.10	0.09	0.07
Red Bluff	0.15	0.13	0.13	0.11	0.10	0.07

Name ____________________

The Long-Distance Connection

Schedule of Rates

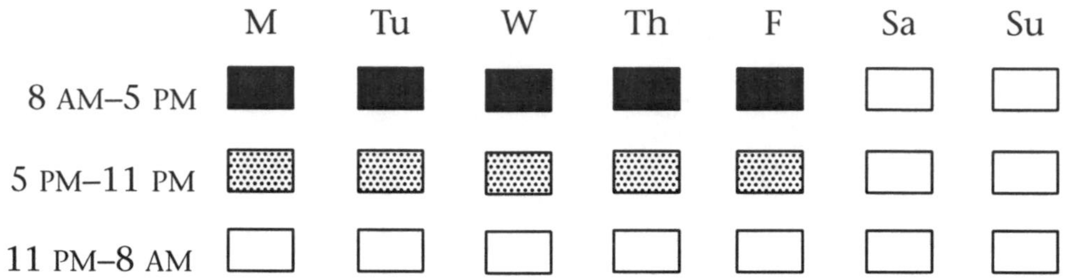

	Day		Evening		Night	
From: Redding **To:**	1st min.	add'l. min.	1st min.	add'l. min.	1st min.	add'l. min.
Alturas	$0.11	$0.07	$0.09	$0.06	$0.07	$0.04
Challenge	0.12	0.10	0.10	0.07	0.08	0.05
Millville	0.13	0.11	0.11	0.09	0.08	0.06
Oroville	0.14	0.12	0.13	0.10	0.09	0.07
Red Bluff	0.15	0.13	0.13	0.11	0.10	0.07

1. Fill in the T-tables to compare the phone rates of calls from Redding to Oroville.

Day

min.	price
0	
1	
2	
3	
⋮	
13	
m	

Evening

min.	price
0	
1	
2	
3	
⋮	
13	
m	

Night

min.	price
0	
1	
2	
3	
⋮	
13	
m	

2. Fill in the T-tables to compare the phone rates of calls from Redding to Red Bluff.

Day		Evening		Night	
min.	price	min.	price	min.	price
0		0		0	
1		1		1	
2		2		2	
3		3		3	
.		.		.	
.		.		.	
13		13		13	
m		m		m	

3. Fill in the T-tables to compare the phone rates of calls from Redding to Challenge.

Day		Evening		Night	
min.	price	min.	price	min.	price
0		0		0	
1		1		1	
2		2		2	
3		3		3	
.		.		.	
.		.		.	
13		13		13	
m		m		m	

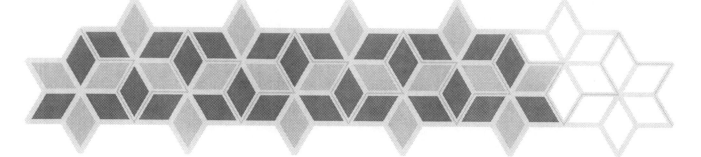

9 A Nutty Function

Objective
Students will use carriage bolts to study slope and intercept.

Materials
Transparency Master 9: A Nutty Function

Worksheet 9: A Nutty Function (two copies if homework is assigned) (1 page)

Transparency of graph from Worksheet 9 (optional)

One carriage bolt and nut (see illustrations) per student

Millimeter rulers

Permanent markers (optional)

Time Required
1 to 2 class periods

Journal Prompts
• How is algebra related to a carriage bolt? In your explanation, include the terms "slope" and "intercept."

• Explain why occasionally your measurements vary after every five turns.

Assessment
Check student graphs, T-tables, and formulas for accuracy.

Homework

Have students construct a graph and T-table and find a formula for a bolt they may have at home. You may wish to issue a nut and bolt to each student to use for homework. The assigned bolts can be identical, or you may choose to use different types. If homework is assigned, each student will require a second copy of Worksheet 9.

Extensions

You can have students analyze bolts that have no "shoulder." These types of bolts have an intercept of 0 and result in the graph of a direct variation. From a direct variation, ratios and proportions can be illustrated, for example, $\frac{4 \text{ mm}}{5 \text{ turns}} = \frac{x \text{ mm}}{15 \text{ turns}}$.

Suggested Teaching Procedure

1. Students can work individually or in pairs for this activity. Tell them that algebra can be found anywhere, even in some pretty "nutty" places! Pass out Worksheet 9: A Nutty Function, one carriage bolt and nut per student or pair, and millimeter rulers. The students should label their graphs as indicated. Set a carriage bolt on the overhead projector or use Transparency Master 9 in place of a physical demonstration. If you use an

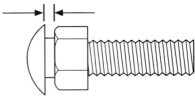

actual bolt, twist the nut until it will go no farther and measure in millimeters the distance between the head of the bolt and the top of the nut (this unthreaded portion is called the "shoulder"). This distance will probably be 1 or 2 millimeters. Have students follow this same procedure using their own bolts.

Students should record this measurement on their graphs as the intercept of the vertical axis. It may be helpful to model plotting this point on a transparency of the student graph. Ask students why this situation represents the vertical intercept or step 0. They should see that it is the starting point for the nut before it is twisted off the bolt.

2. Instruct students to turn the nut five complete turns. You can model this on the overhead projector with your bolt or refer to Transparency Master 9. The students may find it helpful to mark the nut and the head of the bolt with a line from a permanent marker so that they can tell

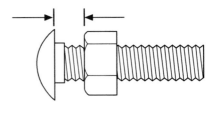

when they have made one complete turn. Measure the distance again.

Record the second measurement on the graph. You may also wish to have students construct and maintain a T-table to record turns and measurements.

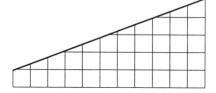

3. Repeat procedure step 2 (five turns, measure, graph) until the nut falls off the bolt.

4. Have students connect the points on their graphs and try to find a formula. This can lead to difficulties. First, since each measurement is made after five turns, students need to divide to find the slope. If the nut moves 7 millimeters after every 5 turns, the slope would be $\frac{7}{5} = 1.4$. You may choose to introduce the "rise over run" definition of slope at this time. Second, the slope may appear to vary. For example, the difference between measurements may vary between 6 and 7 millimeters due to rounding to the nearest millimeter. In this situation, it is recommended that students find the average of the differences before calculating the slope.

5. This activity provides an excellent physical illustration of slope and intercept. After students have graphed the function, take one of their graphs and cut it along the function line and the graph boundaries as shown at right.

Next, tape the wide end of the graph to a pencil and wind the graph around the pencil.

The result is a model of the original bolt. The tapering slope of the function represents the threads of the bolt, and the gap (intercept) at the end near the eraser is the shoulder. Students are welcome to cut out and wrap their graphs too.

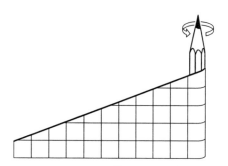

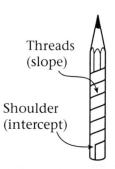

Threads (slope)

Shoulder (intercept)

6. For homework, you may wish to have students analyze the **algebra** in a nut and bolt from home. Bolts that do not have a shoulder **will** have a zero intercept and produce a direct variation. Direct variation **can** then be used to illustrate ratio and proportion.

You can also provide homework by buying different size bolts for the classroom activity. Afterward, students can exchange bolts and take a different one home for homework.

Measure in millimeters:

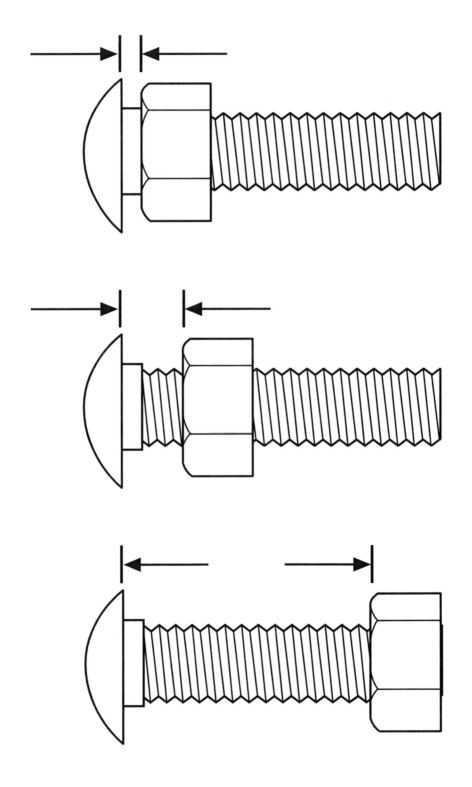

Name _____

A Nutty Function

Find the algebra in a bolt!

Twist the nut onto the bolt as far as it will go. Measure in millimeters the distance between the head of the bolt and the top of the nut. Record this measurement on your graph as the intercept of the vertical axis.

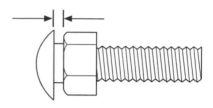

Now turn the nut five complete turns. Measure the distance again. Record the measurement on your graph.

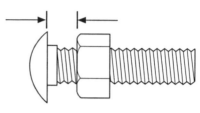

Repeat this process (five turns, measure, graph) until the nut falls off the bolt.

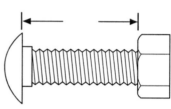

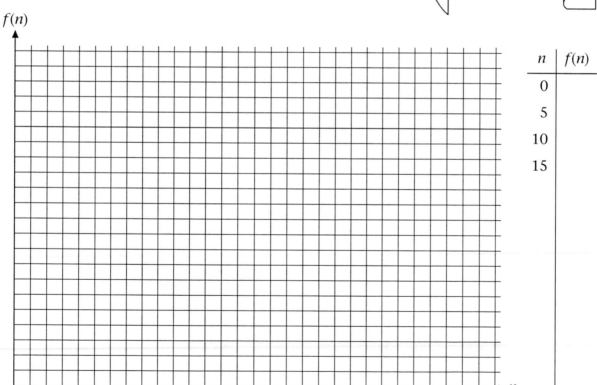

n	f(n)
0	
5	
10	
15	

$f(n)$

Distance in millimeters (count by 4's)

Number of turns (count by 1's)

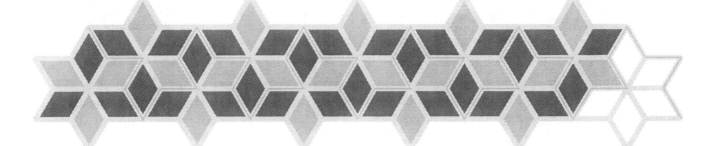

10 The Great Yo-Yo Festival

Objective	Students will study the standard form of a linear equation, $y = mx + b$.
Materials	Transparency Master 10: The Great Yo-Yo Festival Worksheet 10: The Great Yo-Yo Festival (1 page) Homework 10: The Great Yo-Yo Festival (2 pages) Graph paper (optional)
Time Required	1 class period
Journal Prompts	• One year at the yo-yo festival, there were 98 yo-yos. The box of spares held 14 yo-yos. How many experts might there have been, and how many main yo-yos might they have brought? • One year there were 59 total yo-yos, 8 experts, and 15 yo-yos in the box of spares. The judge wants to investigate a potential problem. What did the judge notice?
Assessment	Check student graphs, T-tables, and formulas for accuracy.
Homework	Homework 10: The Great Yo-Yo Festival

Extensions

The first journal prompt, and others like it, can be used as an investigation. Students could be shown how to use a T-table to find possible solutions:

y	$=$	m	$\cdot$	x	$+$	b
98		84	$\cdot$	1	$+$	14
98		42	$\cdot$	2	$+$	14
98		28	$\cdot$	3	$+$	14
98		?	$\cdot$	4	$+$	14
98		?	$\cdot$	?	$+$	14

Suggested Teaching Procedure

1. Students can work individually or in groups. Display Transparency Master 10: The Great Yo-Yo Festival. Use a sheet of paper to hide the transparency. Pass out Worksheet 10 to the students.

2. Display only the first problem on the transparency and ask students to solve it. Most students will have little difficulty, and many will have become accustomed to doing multiplication before addition by this time: $y = 5 \cdot 27 + 34 = 169$. (Be aware that some students may need to be introduced to the multiplication dot symbol.)

3. Ask students what the y represents in this first problem. They should see that it represents the number of "yo-yos."

4. Display the second problem. Here students will be solving for the intercept. Some will use a guess-and-check method. Ask them what guesses they took and why they chose that number. For example, if they discovered that 10 gave the result of 142, ask why they didn't try 5 next. Some will answer that they knew they needed a larger number. On the other hand, several students may choose to use a more formal approach and work backward. Let these students explain this approach to the class. It is much faster than the guess-and-check method and the class will appreciate and retain the approach better if it comes from the discovery of a peer. You should use this opportunity to translate working backward into the formal algebraic solution as follows:

Working Backward

$$154 = 6 \cdot 22 + b$$
$$154 = 132 + b \qquad \text{Find the number of yo-yos with experts.}$$
$$\underline{-132 \qquad -132} \qquad \text{Take away from the total the number of yo-yos with experts.}$$
$$22 = b \qquad \text{The number of yo-yos in the box.}$$

5. Ask students what the *b* stands for. They should see that it represents the yo-yos in the "box." It should be stated that *b* is the intercept or starting value. This can be made more apparent if students write the formula $y = 6x + 22$ and graph it. However, you may want to reserve graphing until after students have completed their worksheets.

6. Display the third problem. Again, most students will use a guess-and-check method, and in this problem, they may need to make more guesses than they did in the second problem. Students should share their strategies with the class. Be sure to include the explanations of students who used more formal strategies. Show how working backward translates into the algebraic solution:

Working Backward

$$182 = m \cdot 31 + 58$$
$$\underline{-58 \qquad\qquad -58} \qquad \text{Take away the number of yo-yos in the box.}$$
$$\frac{124}{31} = \frac{m \cdot 31}{31} \qquad\quad \text{You know 124 yo-yos came with experts.}$$
$$\qquad\qquad\qquad \text{Divide by the number of experts.}$$
$$4 = m \qquad\qquad \text{The number of yo-yos with each expert.}$$

7. Ask the students what the *m* represents. They should say it stands for "main yo-yos." It should be stated that *m* is the slope or the number added with each expert. Again, writing the formula $y = 4x + 58$ and graphing it can help illustrate that *m* is the slope.

8. Next, ask students to solve the fourth problem. Here they solve for *x*, which they should see stands for the number of "experts." Most will still use guess-and-check, but you may find that more of the students are willing to try the formal algebraic approach:

$$214 = 7 \cdot x + 46$$
$$\underline{-46 \qquad\qquad -46} \qquad \text{Take away the number of yo-yos in the box.}$$
$$\frac{168}{7} = \frac{7 \cdot x}{7} \qquad\qquad \text{Divide by the number of main yo-yos.}$$
$$24 = x \qquad\qquad \text{The number of experts.}$$

(This is a good time to show students why the symbol $\times$ is not used to show multiplication in algebra: $214 = 7 \times x + 46$.)

9. Ask students what the *x* represents. They should say it represents the number of "experts," and they will probably ask why not use an "*e*." You can tell them either the authors couldn't spell well or that it really doesn't matter what letter you use in algebra to solve a problem. We prefer the latter explanation!

10. Now direct your students' attention to the fifth problem. Ask them if they can answer it. Most students will say there is not enough information, although some may realize that there is enough to construct a T-table, graph, and formula. If students do not see this, ask them how many yo-yos there will be if only one expert comes. That is 40. What if two experts come? That would be 48. Continue in this manner until students see the pattern, then ask them to find the formula. Some students will see it without a T-table or graph, but you may ask them to construct either one or both. This is strongly suggested since the fifth problem is similar to the homework.

11. Lastly, ask students to translate this function into words:

$$y = m \cdot x + b$$

Let them share their responses with the class. Students should write something similar to this: "The number of total yo-yos is equal to the number of main yo-yos times the number of experts plus the spare yo-yos in the box." With algebra students, it is wise to define $y = mx + b$ as the generic slope-intercept form of a linear equation.

12. Assign Homework 10: The Great Yo-Yo Festival.

Answer Key

Homework 10: The Great Yo-Yo Festival

Questions 1–7 are based on the formula $y = 4x + 12$.

Questions 8–16 are based on the formula $y = 9x + 20$.

1. 40
2. 52
3. 72
4. 120
5. 8
6. 1
7. 18

8. 119
9. 92
10. 164
11. 1316
12. 2
13. 0
14. 10
15. 2324
16. about 109

At the annual yo-yo festival, all the yo-yo experts bring their favorite models. The festival also provides a box of spare yo-yos.

1. Last year, there were 34 yo-yos in the box of spares. There were 27 experts, and each one brought 5 yo-yos. How many yo-yos were there altogether?

$$y = 5 \cdot 27 + 34$$

2. This year, each expert brought 6 yo-yos. The number of experts attending was 22. In all, there were 154 yo-yos. How many spare yo-yos were in the box?

$$154 = 6 \cdot 22 + b$$

3. Two years ago, there were 182 yo-yos altogether, and 58 were in the box of spares. How many main yo-yos did each of the 31 experts bring?

$$182 = m \cdot 31 + 58$$

4. Three years ago, there were 214 yo-yos altogether. The box contained 46 spares. Each expert was told to bring 7 main yo-yos. How many experts attended?

$$214 = 7 \cdot x + 46$$

5. In the future, all experts are to bring 8 main yo-yos. The box of spares will have 32 yo-yos. How many yo-yos will there be altogether?

$$y = \underline{\quad} \cdot \underline{\quad} + \underline{\quad}$$

Name _____

The Great Yo-Yo Festival

At the annual yo-yo festival, all the yo-yo experts bring their favorite models. The festival also provides a box of spare yo-yos.

1. Last year, there were 34 yo-yos in the box of spares. There were 27 experts, and each one brought 5 yo-yos. How many yo-yos were there altogether?

$$y = 5 \cdot 27 + 34$$

2. This year, each expert brought 6 yo-yos. The number of experts attending was 22. In all, there were 154 yo-yos. How many spare yo-yos were in the box?

$$154 = 6 \cdot 22 + b$$

3. Two years ago, there were 182 yo-yos altogether, and 58 were in the box of spares. How many main yo-yos did each of the 31 experts bring?

$$182 = m \cdot 31 + 58$$

4. Three years ago, there were 214 yo-yos altogether. The box contained 46 spares. Each expert was told to bring 7 main yo-yos. How many experts attended?

$$214 = 7 \cdot x + 46$$

5. In the future, all experts are to bring 8 main yo-yos. The box of spares will have 32 yo-yos. How many yo-yos will there be altogether?

$$y = \underline{\quad} \cdot \underline{\quad} + \underline{\quad}$$

The Pattern and Function Connection / Copyright © 2001 Key Curriculum Press

Name _____

The Great Yo-Yo Festival

At this year's yo-yo festival, every expert must bring 4 yo-yos. There will
be 12 yo-yos in the box of spares. First complete the T-table and graph
to show how many yo-yos there will be for different numbers of experts.
Write a function in the form $y = mx + b$. Then use the graph and the
function to answer the questions.

1. If 7 experts arrive, how many yo-yos will there be
    at the festival? _____

2. If 10 experts arrive, how many yo-yos will there be
    at the festival? _____

3. If 15 experts arrive, how many yo-yos will there be
    at the festival? _____

4. If 27 experts arrive, how many yo-yos will there be
    at the festival? _____

5. If there are 44 yo-yos at the festival, how many experts are
    attending? _____

6. If there are 16 yo-yos at the festival, how many experts are
    attending? _____

7. If there are 84 yo-yos at the festival, how many
    experts are attending? _____

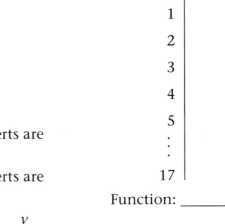

x	y
0	
1	
2	
3	
4	
5	
⋮	⋮
17	

Function: _____

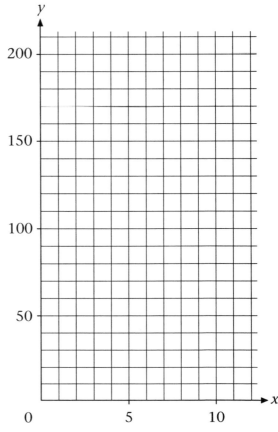

At this year's yo-yo festival, every expert must bring 9 yo-yos. There will be 20 yo-yos in the box of spares. First complete the T-table and graph, and write a function in the form $y = mx + b$. Then use the graph and function to answer the following questions.

8. If 11 experts arrive, how many yo-yos will there be at the festival? _____

9. If 8 experts arrive, how many yo-yos will there be at the festival? _____

10. If 16 experts arrive, how many yo-yos will there be at the festival? _____

11. If 144 experts arrive, how many yo-yos will there be at the festival? _____

12. If there are 38 yo-yos at the festival, how many experts are attending? _____

13. If there are 20 yo-yos at the festival, how many experts are attending? _____

14. If there are 110 yo-yos at the festival, how many experts are attending? _____

15. If 256 experts arrive, how many yo-yos will there be at the festival? _____

16. If there are about 1000 yo-yos at the festival, how many experts are attending? _____

x	y
0	
1	
2	
3	
4	
5	
⋮	
17	

Function: _____

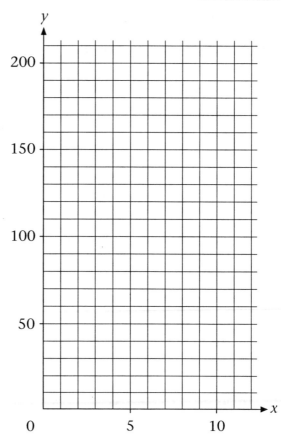

11 The King's Pathway

Objective

Students will design blueprints for pathways made of various polygons. They will construct T-tables and graphs for each polygon and find the formula for each function.

Materials

Transparency Masters 11.1 through 11.5, or task cards of them

Worksheet 11: The King's Pathway (3 pages)

1-centimeter graph paper (or dot paper)

12" × 18" drawing paper

Colored pencils

Rulers

Pattern blocks

Time Required

2 to 5 class periods

Journal Prompts

- Why do the formulas for each individual shape add up to the formula for the entire pathway?

- If you were given the cost of each type of polygon, how would you find the cost of the entire pathway?

- Do your functions have an intercept other than zero? What is it about the design of the pathway that makes an intercept?

- What component of the pathway accounts for the slope of the functions?

Assessment	Ask students to assess each other's projects before submitting them for a grade.

This project can be used as a unit assessment in lieu of, or in addition to, the Linear Functions Test (Appendix C). The King's Pathway may more accurately represent a student's understanding of, and proficiency with, the concepts presented in this book. (Further assessment ideas are listed in the Suggested Teaching Procedure section.)

Homework	With younger students, most of this project should be completed in class. This will allow them the opportunity to ask you questions and discuss with their peers. More-advanced students can work on the project at home. In general, students need about four or five class days to complete this project.

Extensions	Students who finish their projects early can try a second, more advanced pathway. Some students like to explore "The King's Patio." This is a pattern that grows in two directions. Thus many of the resulting formulas are second-degree nonlinear functions.

Suggested Teaching Procedure

1. Students may work individually or in pairs for this activity. Display Transparency Master 11.1: Pathway 1 or a task card of it. Pass out colored pencils and graph paper. Explain that a local amusement park has commissioned you to design the "King's Pathway" through the park. Steps 1, 2, and 3, of a proposed pathway are shown. The pathway is to be built by repeating the pattern for 100 steps.

2. Ask the students to name the three polygons used in the design. They are squares, triangles, and rhombuses. (In the standard pattern block set, the large blue rhombus is called a "parallelogram" since its opposite sides are parallel. However, since all four sides are also equal in length, it is best described as a rhombus. Since there is also a smaller tan rhombus in the set called a "rhombus," we recommend referring to the shapes as the large rhombus and small rhombus when necessary. Avoid referring to them as the "blue" and "tan" since this does not help students develop proper vocabulary.)

3. Ask the students to make a T-table along one side of their graph paper and use the table to determine the function for the number of squares. You can model this procedure by constructing a T-table in the white space on the transparency. Students will see that the sequence of numbers in

the table indicate a slope of 4 and an intercept of 4. Write the function for the number of squares.

Squares $f(n) = 4n + 4$

Have the students graph this function on their graph paper. They should be able to construct a graph with appropriately scaled axes based on what they have learned in previous activities. However, this may be time for a valuable discussion about how the graph should look, how it could be labeled, and what units might be best used on the axes. (You may have to directly instruct students to not place the x-axis at the bottom of the graph paper—room should be left on the y-axis to number it to –6.) Students may each design their graph in different ways. This will make it appear that the slope of the function is different on some graphs. Remind them that slope is not merely "steepness"; it is the amount of increase from one step to another.

Ask students how many squares would be in a 100-step pathway. They should be able to use the function to answer 404 squares.

4. Next have the students make T-tables for the number of rhombuses and triangles in each step of Pathway 1. Graph these two new functions on the same grid as the function for squares. Use a different color pencil for each. Write the functions algebraically.

Rhombuses $f(n) = 2n + 2$
Triangles $f(n) = 4n + 4$

Highlight that the construction company would need to order 202 rhombuses and 404 triangles for the full 100-step pathway.

5. Lastly, the students should construct a T-table for the total number of tiles (squares, rhombuses, and triangles combined) and graph this function on the same grid. Students should notice that the function is the sum of the previous expressions.

Squares $4n + 4$
Rhombuses $2n + 2$
Triangles $4n + 4$
_____
Total $f(n) = 10n + 10$

The entire pathway would have 1010 tiles.

6. Students can work with Pathways 2 through 5 for more practice as needed. Ask them what makes some pathways have a positive intercept, as opposed to an intercept of 0. Ask them what causes a negative intercept. How could a pathway be modified to change its intercept? What determines the slope of the path's function? Understanding the answers to these questions will help students as they design their own pathways.

---

7. When students are proficient with writing functions for a given path, remind them that the amusement park has commissioned them to design a unique pathway. Distribute Worksheet 11: The King's Pathway. For this project, students will draw a blueprint that shows steps 1, 2, and 3 of their own design. They will need T-tables and functions for each shape of tile they use and each function will need to be graphed. Lastly, the builder needs to know how many of each tile must be ordered; thus each function must be evaluated at step 100.

8. Allow students time to explore with the pattern blocks and to experiment with different pathway designs. Once they have selected a design, each student should trace around pattern blocks to create the design on drawing paper. They should take care that the three steps of the path are straight, parallel to the top and bottom of the paper, and evenly spaced vertically. In addition to demonstrating an understanding of functions, students are putting measurement skills to work. They will likely need some direction on how to draw their paths neatly.

   Students should draw their T-tables with a ruler, not freehand. For their graphs, you may wish to give students pieces of graph paper to glue or tape onto their projects—1-centimeter grids work well for this project. You may also wish to have students put the T-tables and graphs on the back of the drawing paper or on a separate sheet of paper. This way, the finished paths can be displayed on the wall, and other students can be challenged to find the functions.

   The colors of the T-tables, algebraic functions, and lines on the graphs should match the colors used in the polygon itself. Thus, if a student colors the triangles blue, the same color should be used for the mathematics that refer to the triangles. Students need not be restricted to the colors used in the pattern block set.

9. It will probably take students one class period to design a path. Another period will be spent transferring the path to the large paper. At least one more period will be spent completing the mathematics and coloring the project. Students who finish early can color the backgrounds of their projects to make them look even more presentable, or they can design a second, more advanced pathway.

10. Explain how the park's administration will grade each project. The following method works well. Here, the project is worth a total of 100 points.

    25 points: Accuracy of Measurement

    Are the paths square to and parallel to the edges of the paper?

    Is the vertical spacing between the three steps equal?

    Are the three steps aligned on the left side?

    Are the T-tables and the graph neatly drawn?

25 points: Quality of Presentation

Is the lettering straight, consistent, and neat?

Are colors used consistently?

Is the pathway colored neatly?

Is the pathway visually pleasing to the eye?

Are erasures neat and complete?

50 points: Mathematics

Are the T-tables accurate?

Are the functions correct?

Are the functions graphed correctly?

Is step 100 correctly evaluated?

Although the requirements of this project are stringent, assessment need not be time-consuming for the teacher. The eye can perceive a variation of as little as one degree from a parallel line. Our eye can also discern the presentational quality and neatness of the projects—we know when something looks good to us.

Lastly, there is a trick to evaluating all the complex mathematics of this project. Simply add all the values the student has calculated for step 100 of the individual shapes. If they sum to step 100's value for the total function, the mathematics that got the student that far is correct. If not, it shouldn't take long to track down the error. For example, in Pathway 1:

	Step 100 on the square T-table:	404
	Step 100 on the rhombus T-table:	202
+	Step 100 on the triangle T-table:	404
	Step 100 on the total T-table should be:	1010

There are many advantages to allowing students time to work on a project of this scope. First, it allows you the freedom to move around the room, providing help to individual students. Second, working on this project will do more to cement understanding than an equal amount of time spent on dozens of worksheet problems. With this task, students are fully immersed in the multiple representations of functions. Finally, research shows that students' self-esteem grows as they recognize their own accomplishment. As they work on this project, and develop confidence in themselves, you will see your students exceed their own expectations.

Answer Key

Transparency Masters 11.1–11.5

Pathway 1

Squares	$f(n) = 4n + 4$
Rhombuses	$f(n) = 2n + 2$
Triangles	$f(n) = 4n + 4$
Total	$f(n) = 10n + 10$

Pathway 2

Triangles	$f(n) = 4n - 2$
Rhombuses	$f(n) = 4n - 2$
Trapezoids	$f(n) = 4n - 2$
Total	$f(n) = 12n - 6$

Pathway 3

Squares	$f(n) = 2n + 1$
Large rhombuses	$f(n) = 2n$
Small rhombuses	$f(n) = 2n$
Trapezoids	$f(n) = 2n$
Total	$f(n) = 8n + 1$

Pathway 4

Squares	$f(n) = 4n - 2$
Large rhombuses	$f(n) = 2n$
Small rhombuses	$f(n) = 4n - 2$
Hexagons	$f(n) = 4n - 2$
Total	$f(n) = 14n - 6$

Pathway 5

Squares	$f(n) = 4n + 4$
Large rhombuses	$f(n) = 3n + 3$
Small rhombuses	$f(n) = 2n + 2$
Triangles	$f(n) = 2n + 2$
Hexagons	$f(n) = n + 1$
Total	$f(n) = 12n + 12$

1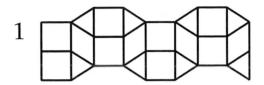

2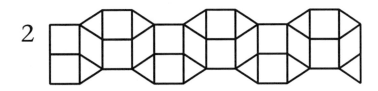

3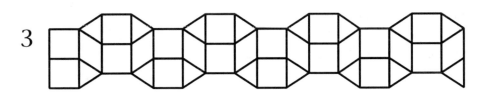

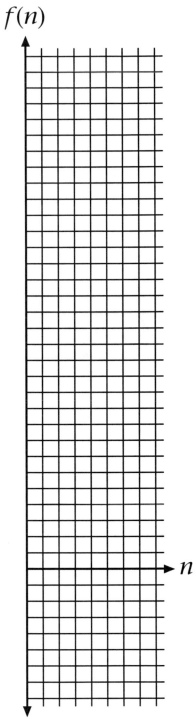

1

2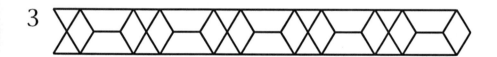

3

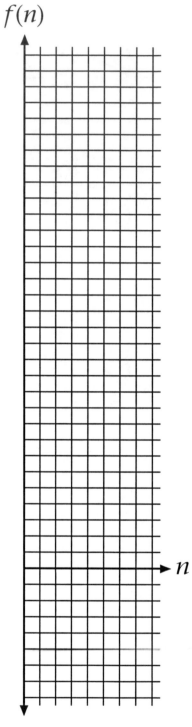

1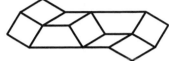

2

3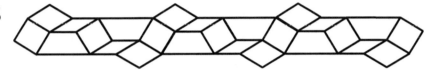

$f(n)$

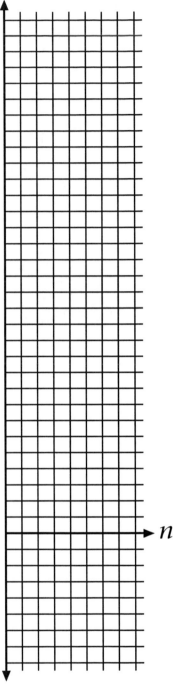

n

1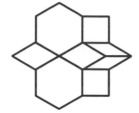

2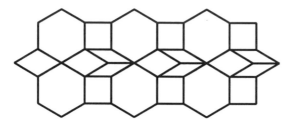

3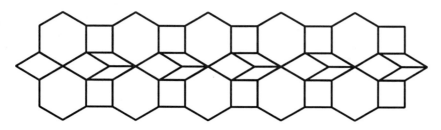

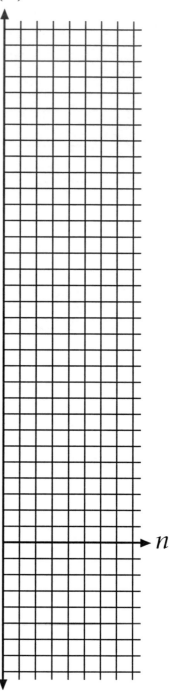

$f(n)$

n

1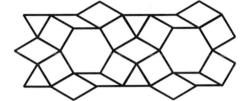

2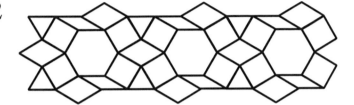

3

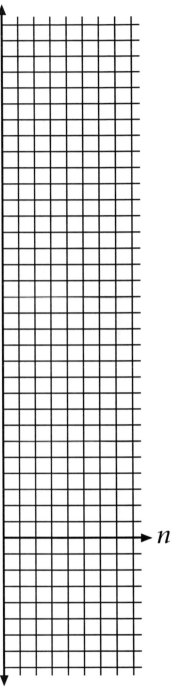

Name _____

The King's Pathway

An amusement park has asked you to design the "King's Pathway" through the park. The pathway is to be built by repeating a pattern for 100 steps.

Use pattern blocks to design a geometric pathway. You must use at least three different shapes to create your design. A complete example is shown at the end of this worksheet.

Here are the names of the shapes used in a set of pattern blocks:

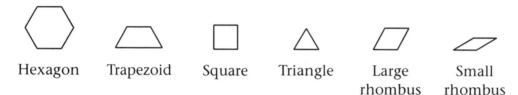

Hexagon   Trapezoid   Square   Triangle   Large rhombus   Small rhombus

1. Begin by finding a paving unit. This is a group of tiles that will connect without spaces.

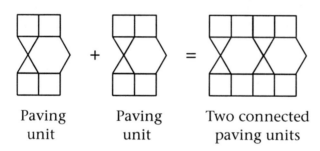

Paving unit   Paving unit   Two connected paving units

2. Next you may wish to construct a starting design that differs from the paving unit. This makes the functions more advanced, which greatly pleases the park's administration.

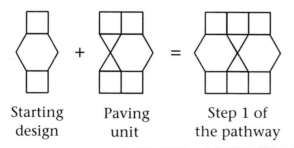

Starting design   Paving unit   Step 1 of the pathway

3. On drawing paper, trace pattern blocks to show your pathway for steps 1, 2, and 3. Make your drawings as neatly as possible and label each step. Color the pathway using a different color for each type of shape. You don't have to use the same colors that the pattern blocks use.

4. Construct a T-table for each shape you have used and for the total number of pattern blocks. Label each T-table with its correct geometric name. Find the algebraic pattern and write a formula. Calculate how many tiles would be used to create the one-hundredth repetition of your path.

5. Graph the functions for the number of each shape and the total number of pattern blocks. Use colored pencils to match each graph with your drawing of the pathway.

6. You will be graded using the following scale:

Accuracy of Measurement:	25 points
Quality of Presentation:	25 points
Mathematics:	50 points
Total:	100 points

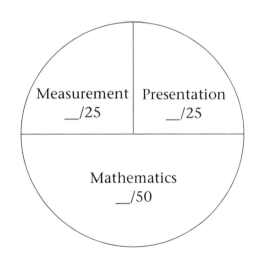

The King's Pathway (Example)

1

2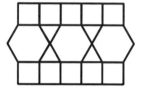

3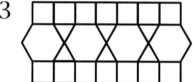

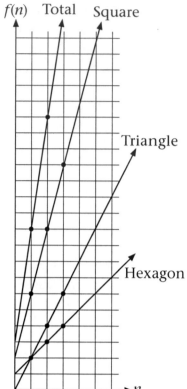

Square			Hexagon			Triangle			Total	
n	$f(n)$		n	$f(n)$		n	$f(n)$		n	$f(n)$
0	2		0	1		0	0		0	3
1	6		1	2		1	2		1	10
2	10		2	3		2	4		2	17
3	14		3	4		3	6		3	24
n	$4n + 2$		n	$n + 1$		n	$2n$		n	$7n + 3$
100	402		100	101		100	200		100	703

Pattern Masters

This appendix contains three items central to the use of this book.

Task Card Instructions (page 88)

As mentioned in the Teaching Patterns and Functions section, task cards can be useful tools for some students. A copy of the Task Card Instructions should be applied to each task card. The instructions tell students the necessary procedure for independently exploring patterns from the task card.

Whether simply laminating photocopies of the Pattern Masters, or creating larger versions with construction paper, you should utilize these instructions. The instructions can be photocopied, cut down to size, and attached to the task card in the available space. The instructions can be glued directly to the card and laminated, or you can recycle a few copies by paper clipping them to only the cards you are using on that day.

If you prefer to work only with transparencies, it will be helpful to have a transparency of the Task Card Instructions. This way, when students are working independently, you can display the instructions for the whole class.

Pattern Masters Key (pages 89–94)

The Pattern Masters Key acts as a quick-glance teacher's answer key for the Pattern Masters. For example, when working with Function A in class, most students will discover a *linear* pattern, based on the number of *tiles*, represented by the function $f(n) = 3n + 1$; this information is succinctly presented in the Pattern Masters Key as

> linear
> tiles
> $f(n) = 3n + 1$

Please note that students may find multiple patterns for some Pattern Masters. In this way, the Pattern Masters Key is not the final word. For example, Function A could also be examined as a linear pattern, based on the perimeter of each figure, represented by the function $f(n) = 2n + 8$.

Pattern Masters (pages 95–146)

A collection of 52 picture and number patterns are presented for use with the activities in this book. The patterns are named Function A to Function ZZ. These Pattern Masters can be used as transparencies and/or made into task cards.

1. Study the pattern.

2. Build step 4.

3. Make a sketch of step 4 and step 5 in your notes. How many units make up each step?

4. Use the pattern you discover to sketch step 10. How many units make up step 10?

5. Describe any number patterns you notice.

6. Suggest other ways to describe how this pattern grows.

Function A

linear
tiles
$f(n) = 3n + 1$

Function B

linear
tiles
$f(n) = 3n + 1$

Function C

linear
toothpicks
$f(n) = 3n$

Function D

linear
tiles
$f(n) = 4n - 3$

Function E

linear
toothpicks
$f(n) = 2n + 2$

Function F

linear
tiles
$f(n) = 2n + 4$

Function G

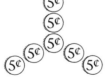

linear
value
$f(n) = 0.15n - 0.10$

Function H

linear
toothpicks
$f(n) = 4n + 1$

Function I

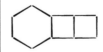

linear
toothpicks
$f(n) = 3n + 3$

Function J	Function K	Function L
linear tiles $f(n) = 4n$	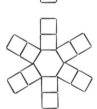 linear toothpicks $f(n) = 18n - 12$	linear tiles $f(n) = n$

Function M	Function N	Function O
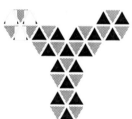 linear tiles $f(n) = 18n - 17$	linear time $f(n) = 2.5n$	nonlinear dots on top $f(n) = 2^{n-1}$

Function P	Function Q	Function R
 nonlinear toothpicks Odd-numbered steps follow $f(n) = 2.5n + 0.5$ while even steps follow $f(n) = 2.5n.$	 linear tiles $f(n) = 4n + 2$	nonlinear tiles $f(n) = n^2$

The Pattern and Function Connection

Function S

linear
toothpicks
$f(n) = 3n + 3$

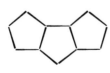

Function T

linear
toothpicks
$f(n) = 3n + 4$

Function U

linear
toothpicks
$f(n) = 3n + 5$

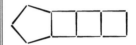

Function V

linear
toothpicks
$f(n) = 3n + 6$

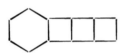

Function W

linear
toothpicks
$f(n) = 2n + 1$

Function X

linear
toothpicks
$f(n) = 3n + 1$

Function Y

linear
toothpicks
$f(n) = 4n + 1$

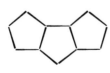

Function Z

linear
toothpicks
$f(n) = 5n + 1$

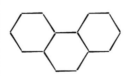

Function AA

nonlinear
toothpicks
$f(n) = \frac{3n^2 + 3n}{2}$

Function BB

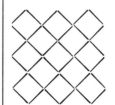

nonlinear
toothpicks
$f(n) = 4n^2$

Function CC

nonlinear
tiles
$f(n) = 2n^2 - 2n + 1$

Function DD

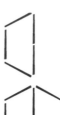

nonlinear
toothpicks
Odd-numbered
steps follow
$f(n) = 3.5n + 1.5$
while even
steps follow
$f(n) = 3.5n + 1.$

Function EE

linear
tiles
$f(n) = 3n$

Function FF

linear
tiles
$f(n) = 5n + 1$

Function GG

nonlinear
tiles
$f(n) = n^2 + n$

Function HH

nonlinear
tiles
$f(n) = n^2$

Function II

nonlinear
toothpicks
$f(n) = 2n^2 + 2n$

Function JJ

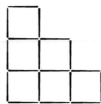
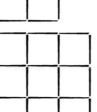

nonlinear
toothpicks
$f(n) = n^2 + 3n$

Function KK

linear
tiles

$f(n) = 3n - 2$

Function LL

linear
toothpicks

$f(n) = 7n - 2$

Function MM

linear
tiles

$f(n) = 9n - 3$

Function NN

linear
tiles

$f(n) = 12n - 10$

Function OO

linear
tiles

$f(n) = -3n + 10$

Function PP

linear
toothpicks

$f(n) = -4n + 25$

Function QQ

linear
arrows

$f(n) = -4n + 22$

Function RR

linear
tiles

$f(n) = -3n + 19$

Function SS

linear
tiles

$f(n) = \frac{3}{2}n + \frac{9}{2}$

Function TT

linear
tiles
$f(n) = \frac{5}{2}n + \frac{5}{2}$

Function UU

linear
value
$f(n) = 0.15n + 0.15$

Function VV

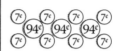

linear
value
$f(n) = 1.08n + 0.14$

Function WW

linear
tiles
$f(n) = \frac{1}{3}n + \frac{2}{3}$

Function XX

$4.38

$5.67

$6.96

linear
value
$f(n) = 1.29n + 3.09$

Function YY

linear
tiles
$f(n) = \frac{3}{2}n - \frac{1}{2}$

Function ZZ

linear
tiles
$f(n) = \frac{4}{3}n + \frac{2}{3}$

Function A

1

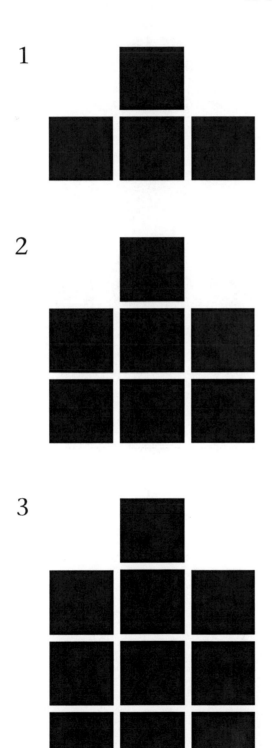

2

3

Function B

1

2

3

Function C

1

2

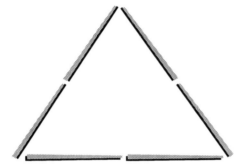

3

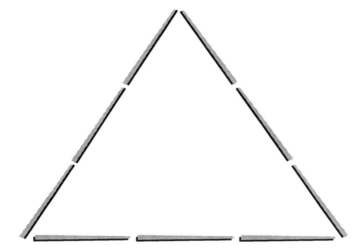

Function D

1

2

3

Function E

1

2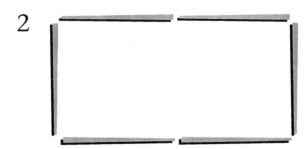

3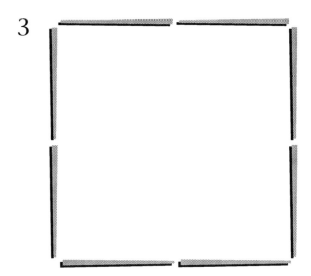

Function F

1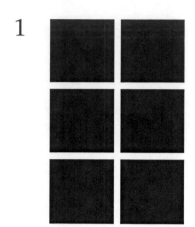

2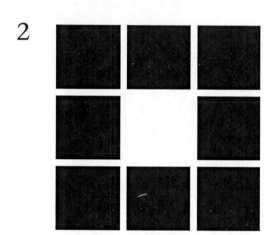

3

Function G

1

2

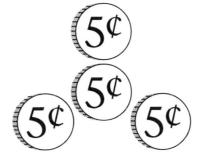

3

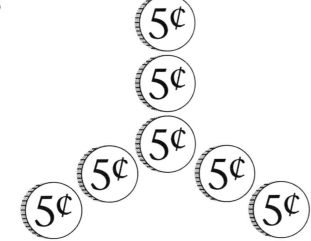

Function H

1

2

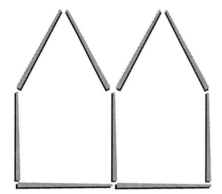

3

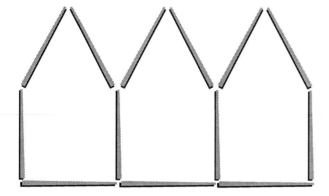

Function I

1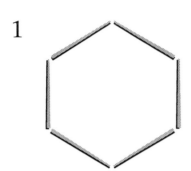

2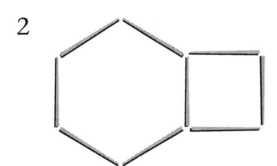

3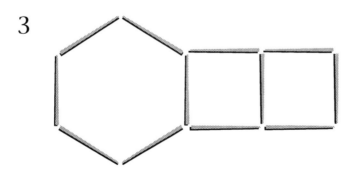

Function J

1

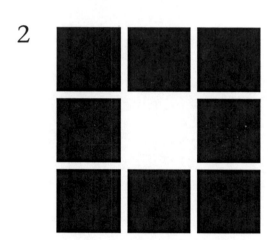

2

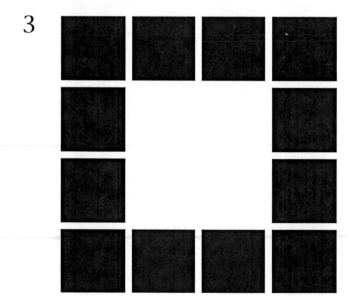

3

Function K

1

2

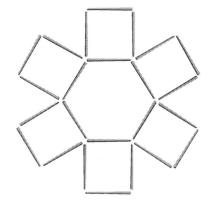

3

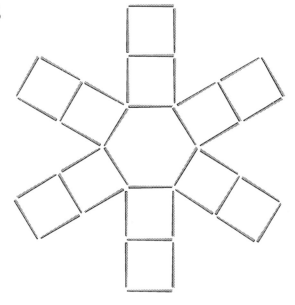

Function L

1

2

3

Function M

1
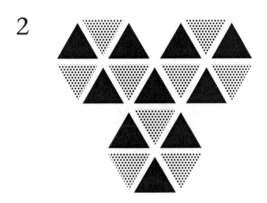

2

3
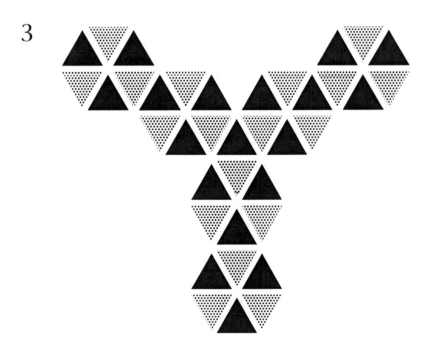

Function N

1

2

3

Function O

1

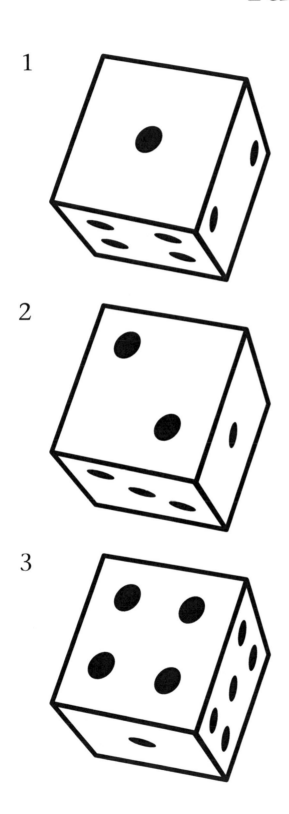

2

3

Function P

1

2
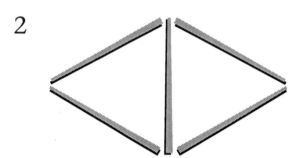

3
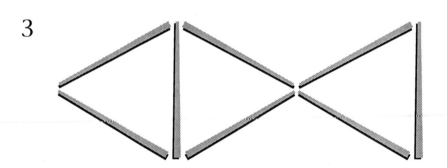

The Pattern and Function Connection / Copyright © 2001 Key Curriculum Press

Function Q

1

2

3

Function R

1

2

3

The Pattern and Function Connection / Copyright © 2001 Key Curriculum Press

Function S

1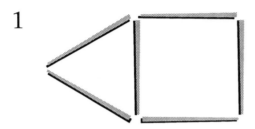

2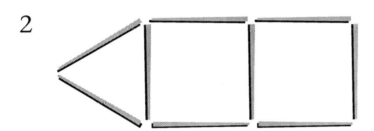

3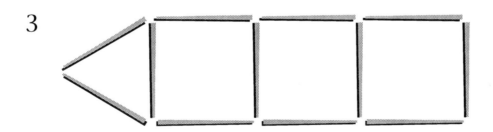

Function T

1

2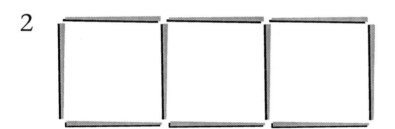

3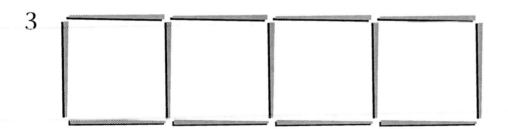

Function U

1
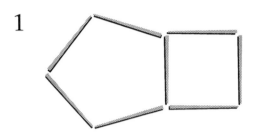

2
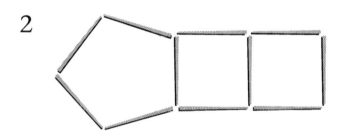

3
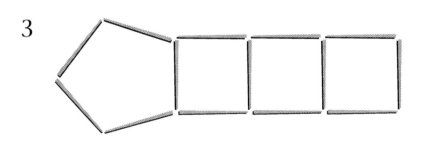

Function V

1
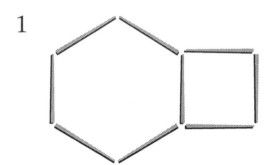

2
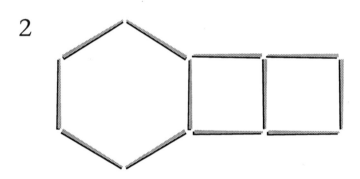

3
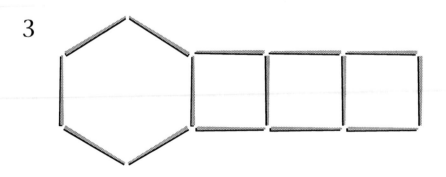

Function W

1

2

3

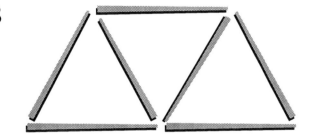

Function X

1

2

3

The Pattern and Function Connection / Copyright © 2001 Key Curriculum Press

Function Y

1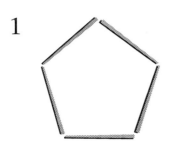

2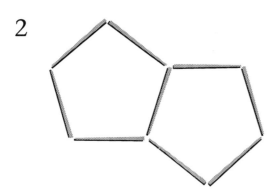

3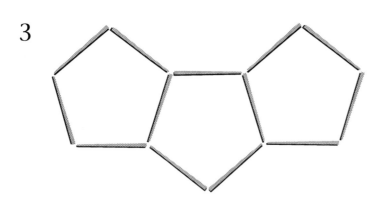

Function Z

1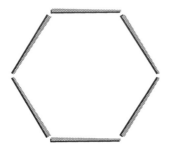

2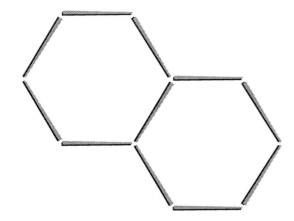

3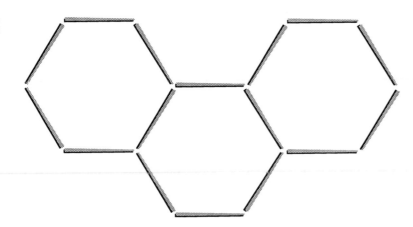

The Pattern and Function Connection / Copyright © 2001 Key Curriculum Press

Function AA

1

2

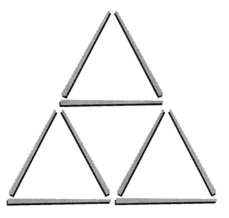

3

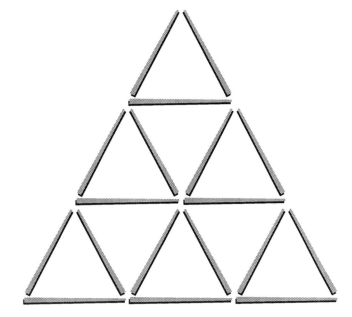

Function BB

1

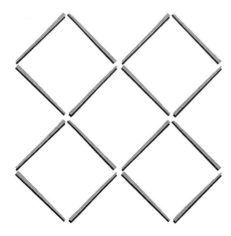

2

3

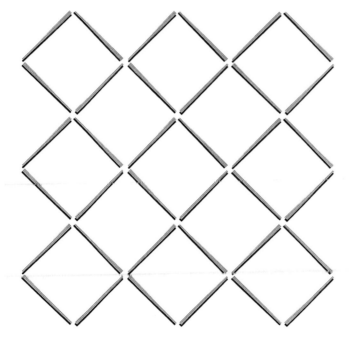

Function CC

1

2

3

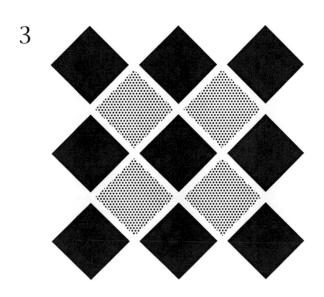

Function DD

1

2

3

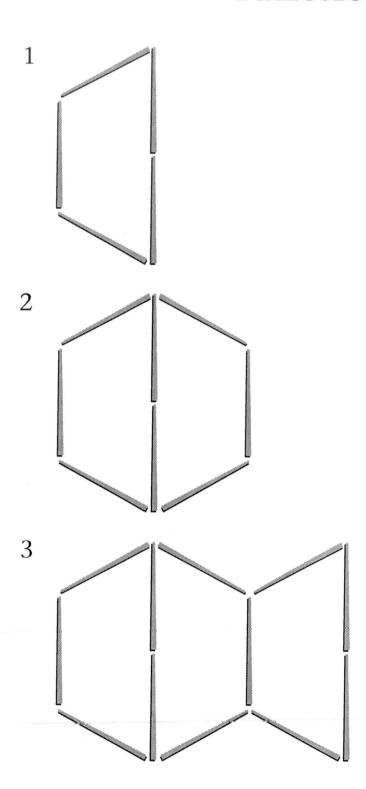

Function EE

1

2

3

Function FF

1

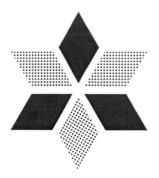

2

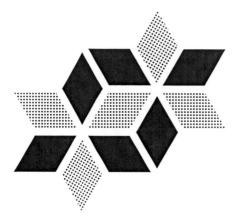

3

The Pattern and Function Connection / Copyright © 2001 Key Curriculum Press

Function GG

1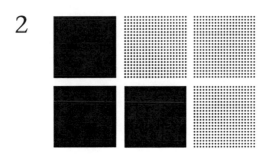

2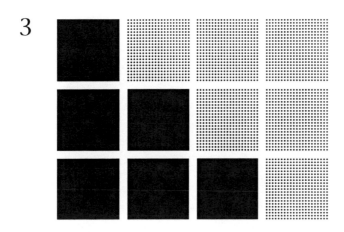

3

Function HH

1

2

3

The Pattern and Function Connection / Copyright © 2001 Key Curriculum Press

Function II

1

2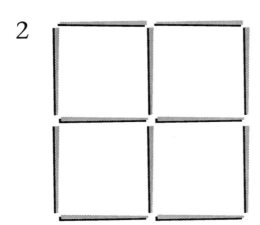

3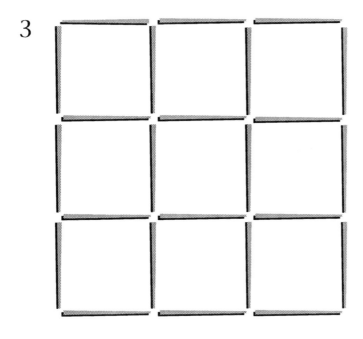

Function JJ

1

2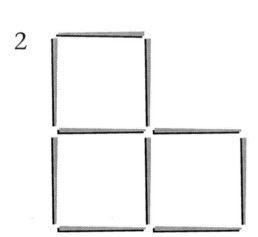

3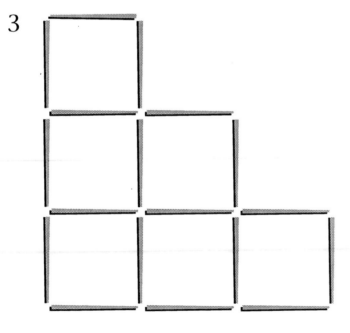

The Pattern and Function Connection / Copyright © 2001 Key Curriculum Press

Function KK

1

2

3

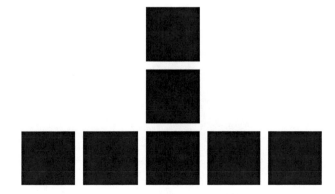

Function LL

1

2

3

Function MM

1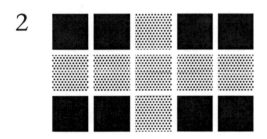

2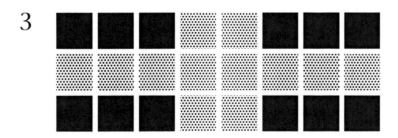

3

Function NN

1

2

3

Function OO

1

2

3

Function PP

1

2

3

Function QQ

1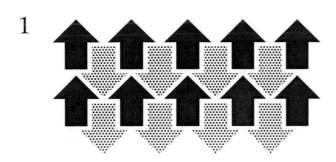

2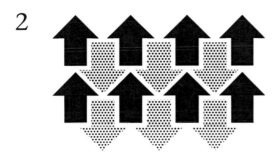

3

Function RR

1

2

3

Function SS

1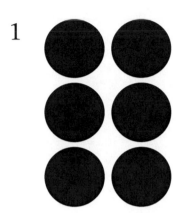

2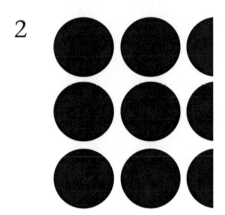

3

Function TT

1

2

3

Function UU

1   (15¢)(15¢)

2   (15¢)(15¢)(15¢)

3   (15¢)(15¢)(15¢)(15¢)

Function VV

1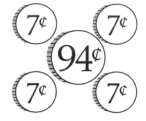

2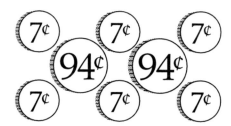

3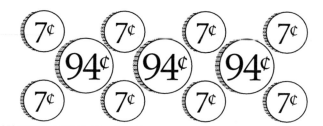

Function WW

1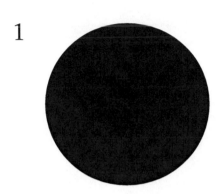

2

3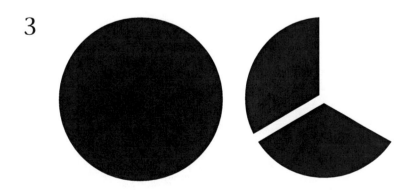

Function XX

1 **$4.38**

2 **$5.67**

3 **$6.96**

Function YY

1

2

3

4

5

Function ZZ

1

2

3

4

5

More Problem-Solving Activities

This appendix contains four additional problems to support the activities in this book. Each problem can be used as a transparency master or as a worksheet (one page each). The problems may also be used as journal prompts for students to practice their new skills or to generate some rich class discussions. The first three problems can be used with beginning students as problems-of-the-day or problems-of-the-week. The last problem, on long-distance phone rates, is more complex and should be used with more advanced students.

All four problems offer students the opportunity to use their knowledge of functions in analyzing real-world problems. Students' conceptual understanding and computational expertise improve when problems are presented in context as they are here. These problems can serve as a springboard for you to design similar problems, or for the students to discover similar problems in their own daily lives. For example, some students may have parents who earn a commission in addition to a flat pay rate; the flat rate is an intercept, and the commission is a slope. A comparison between the Celsius and Fahrenheit temperature scales will also produce a series of points along a line that is the conversion formula for these two scales; if the vertical axis represents the Fahrenheit scale, the intercept is 32, the freezing temperature of water in that scale.

For each problem, encourage students to construct T-tables and graphs, and to find formulas to support their conclusions. Following are suggested teaching procedures and the problem solutions.

The Air Ways Helicopter Company

This problem states that the company spent $11 million building their first helicopter. Subsequent units will cost $3 million apiece. Ask the students why the first unit costs more than the other ones. Discuss that this could be due to the cost of research or the costs of setting up production. Ask the students what variable they wish to use in this function. Some may suggest using an h for "helicopters." Then $f(h)$ is the total cost of producing h helicopters. To begin answering the first question, have students build a T-table for 0 to 3 helicopters. In the T-table, the extra $8 million is the intercept. The $3 million increase per unit is the slope. This results in the cost function $f(h) = 3h + 8$. (It should be noted that the 3 and the 8 represent millions of dollars in this context.) The answer to the second question is that the company can build exactly 13 helicopters for $47 million. For the final question, students need to compare $f(h) = 3h + 8$ to $f(h) = 5.5h$, the income function. By extending tables or by graphing, students will see that the company must sell at least 4 helicopters to make a profit. With 4 helicopters, the cost is $20 million, and the income is $22 million, for a profit of $2 million.

The New Heights Ladder Company

This is really two problems in one. The first part of the activity is similar to a toothpick pattern. The students count the number of feet of wood used to construct the ladders. This results in the formula $f(m) = 3m + 8$, where $f(m)$ is the number of feet used to make the ladder and m is the model number, so the Model 13 ladder will require 47 feet of lumber. Next the students look at the linear relationship for the prices of the ladders. The formula for this pattern is $f(m) = 3.50m + 6.45$, where $f(m)$ is the cost of the ladder and m is the model number. The Model 13 ladder will cost $51.95.

Al's Tune-up Special

This may be the first instance when students are presented with data that is not consecutive. That is, a tune-up price is given for a four-cylinder engine and a six-cylinder engine, but not for a five-cylinder engine. Ask the students why the data is presented this way. Discuss that it is due to the fact that in this real-life context, five-cylinder cars are not common.

In this problem, the price increase from a four-cylinder car tune-up to a six-cylinder car tune-up is $15.00. Therefore, an increase of only one cylinder would be half of that, or $7.50. This is the slope of the function. Ask students to set up a T-table or to construct a graph—either will show that the intercept is $39.95. Thus the formula is $f(c) = 7.50c + 39.95$, where $f(c)$ is the total price of the tune-up and c is the number of cylinders. So a fair price for a twelve-cylinder car would be $129.95; for a five-cylinder car, $77.45; for a three-cylinder motorcycle, $62.45; and for a one-cylinder lawn mower, $47.45.

Connecting with a Good Deal

This activity requires students to compare three different functions. Each function represents a separate calling plan. The first plan has a slope of $0.19 and an intercept of $65. The second plan has a zero intercept but has a slope of $0.25. The third plan has an intercept of $249 and a slope of zero. Thus the three formulas are

$$f(m) = 0.19m + 65$$
$$f(m) = 0.25m$$
$$f(m) = 249$$

In each formula, $f(m)$ represents the cost of a phone service for a total of m minutes. Students should be careful to note that there are decimals in the slope but not in the intercepts; this could lead to some problems with addition when computing answers. This problem could be solved by T-table analysis, by graphing, or by simply solving the algebraic formulas. A comparison of all three methods helps students make connections among the various representations. For Tracy's business, Plan 2 is the best buy. More generally, for 0 to 996 minutes of calls, Plan 2 is the least expensive. For 996 minutes or more, Plan 3 is cheapest. At no time will the customer save money by choosing Plan 1.

The Air Ways Helicopter Company

The Air Ways Helicopter Company is planning to produce its latest model, the Air Star. After extensive research, the company has determined that it will cost $11 million to produce the first one. This amount includes the cost of setting up the factory and the machinery. After that, each additional helicopter will cost $3 million to produce.

How much money will the company need to build the helicopters?

How many helicopters can it build for $47 million?

If the company charges $5.5 million for each helicopter, how many must it sell before making a profit?

The Pattern and Function Connection / Copyright © 2001 Key Curriculum Press

The New Heights Ladder Company

The New Heights Ladder Company uses lumber to make its ladders. Here is how many feet of lumber are needed to make each ladder:

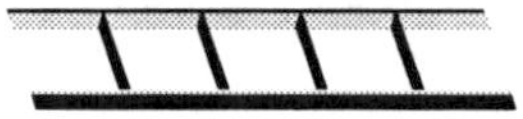

Model 1	Model 2	Model 3
11 feet	14 feet	17 feet

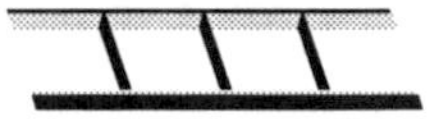

How many feet of lumber are required to make the new Model 13 ladder?

The New Heights Ladder Company prices its three shortest models as shown below.

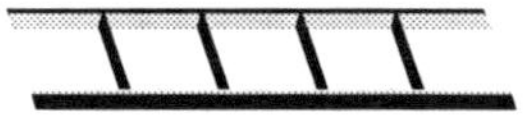

Model 1	Model 2	Model 3
$9.95	$13.45	$16.95

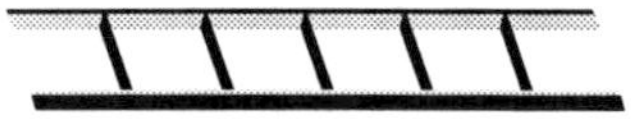

What should the price be for the Model 13 ladder?

Al's Tune-up Special

In the window of Al's Auto Repair, the following sign shows what Al charges for tune-ups:

A customer brings in a twelve-cylinder Ferrari. What would be a fair price?

A second customer brings in a five-cylinder Volkswagen. What should Al charge to be fair?

A third customer brings in a three-cylinder motorcycle. Al agrees to tune it up for the same rate he charges on autos. What should Al charge?

The last customer of the day brings in his one-cylinder lawn mower. Al agrees to tune it up for the same rate he charges on cars. What will the customer have to pay?

Connecting with a Good Deal

Tracy's business is trying to decide which long-distance phone plan is least expensive. Typically, the business makes about 14 hours of long-distance phone calls per month.

Which of these three plans is the best buy for Tracy's business?

Plan 1:

Long-distance service for just 19¢ a minute*

*Additional monthly service charge of $65 applies

Plan 2:

NO MONTHLY SERVICE CHARGE!
Just 25¢ a minute on long-distance calls!

Plan 3:

Unlimited long-distance service for only
$249 a month

Is there a plan that is always the cheapest for any volume of long-distance calls? Justify your answer with T-tables, graphs, or formulas.

Linear Functions Test

This appendix contains an optional paper-and-pencil test (two pages) that can be used as a culminating assessment for the unit. The Linear Functions Test will help support the informal observations you made during each activity.

Answer Key

Linear Functions Test

Questions 1 through 5 are based on the formula $f(n) = 5n + 1$.

2.

n	$f(n)$
1	6
2	11
3	16
4	21
5	26

4.

n	$f(n)$
10	51
25	126
57	286

5. $f(n) = 5n + 1$

6. 5

7. 1

8. $f(m) = 0.11m + 0.02$

9. $f(m) = 0.09m + 0.02$

10. $f(m) = 0.06m + 0.02$

1. Study this pattern. Count the number of tiles in each step. Then complete problems 2 through 7.

2. Begin this T-table:

n	$f(n)$
1	
2	
3	
4	
5	
$\vdots$	

4. Now find:

10	
25	
57	

5. Find a formula for the pattern.

$f(n) =$

3. Graph the function here:

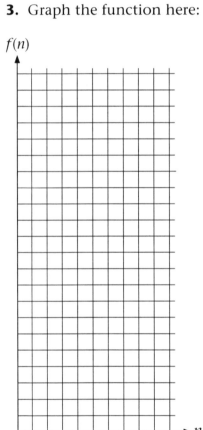

6. What is the slope of this function? _____

7. What is the intercept of this function? _____

Schedule of Rates

	M	Tu	W	Th	F	Sa	Su
8 AM–5 PM	■	■	■	■	■	☐	☐
5 PM–11 PM	▨	▨	▨	▨	▨	☐	☐
11 PM–8 AM	☐	☐	☐	☐	☐	☐	☐

	■ Day		▨ Evening		☐ Night	
From: Redding **To:**	1st min.	add'l. min.	1st min.	add'l. min.	1st min.	add'l. min.
Alturas	$0.11	$0.07	$0.09	$0.06	$0.07	$0.04
Challenge	0.12	0.10	0.10	0.07	0.08	0.05
Millville	0.13	0.11	0.11	0.09	0.08	0.06
Oroville	0.14	0.12	0.13	0.10	0.09	0.07
Red Bluff	0.15	0.13	0.13	0.11	0.10	0.07

Fill in the T-tables to compare the phone rates of calls from Redding to Millville.

8. Day

min.	price
0	
1	
2	
3	
⋮	
13	
n	

9. Evening

min.	price
0	
1	
2	
3	
⋮	
13	
n	

10. Night

min.	price
0	
1	
2	
3	
⋮	
13	
n	

Graph Paper Masters

As students progress through the activities in this book, they are gradually led from using pre-labeled graphing grids, to labeling the provided grids, to constructing their own grids on blank graph paper. This appendix contains four blackline masters of square graph and dot graph paper for use with the later activities.

Please note that some students prefer to use dot graph paper rather than square graph paper; dot graph paper helps these students identify the points formed by intersecting grid lines. For the activities that require graph paper, you may use either square graph or dot graph paper, or both, whichever your students prefer.

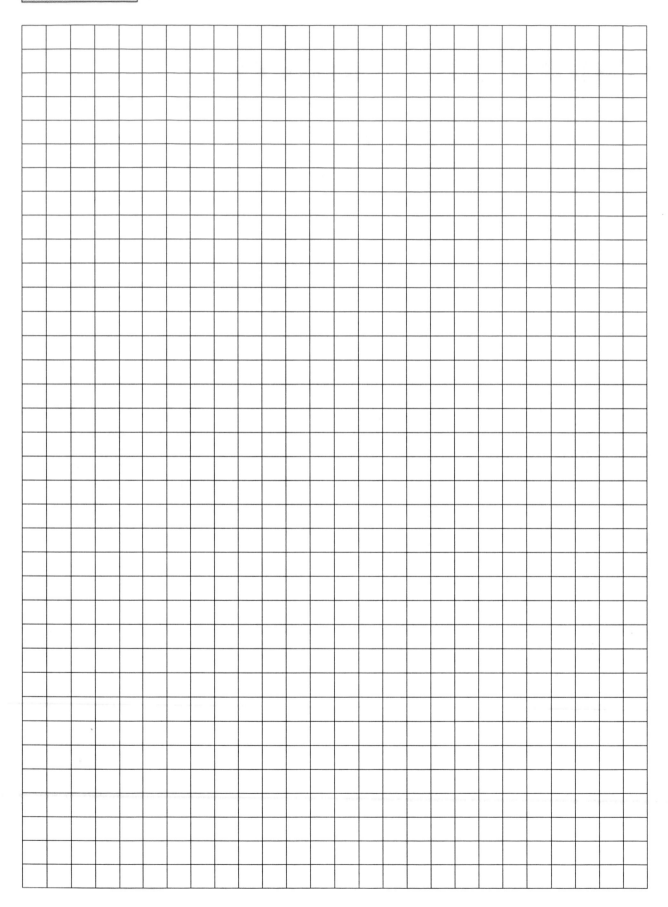

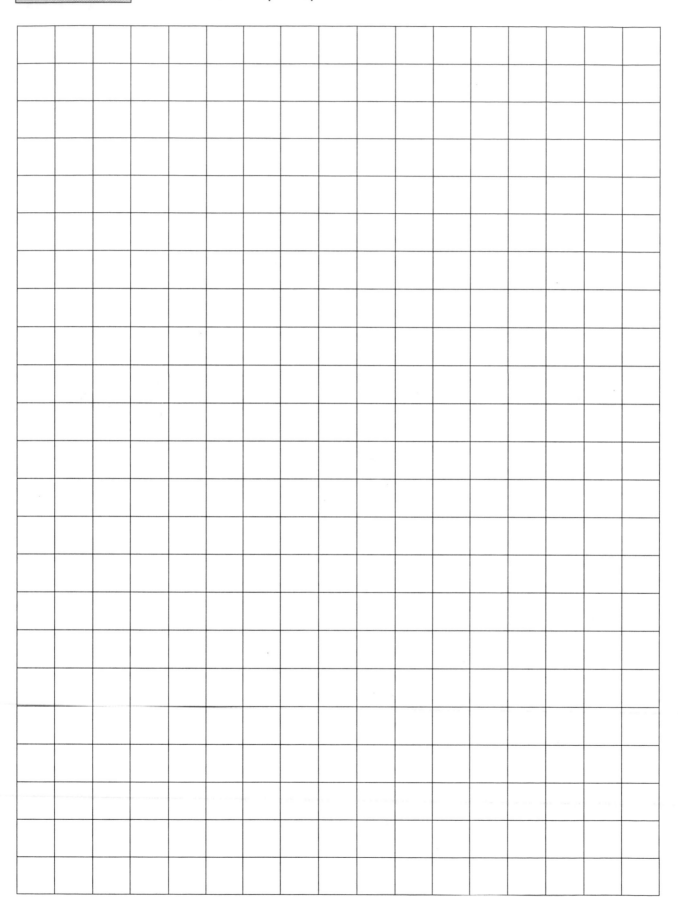